BARRON'S

Painless Pre-Algebra

THIRD EDITION

Amy Stahl, M.S.Ed.

Published by Kaplan, Inc., d/b/a Barron's Educational Series
750 Third Avenue
New York, NY 10017
www.barronseduc.com

ISBN: 978-1-5062-7315-0

10 9 8 7 6 5 4 3 2

Kaplan, Inc., d/b/a Barron's Educational Series print books are available at special quantity discounts to use for sales promotions, employee premiums, or educational purposes. For more information or to purchase books, please call the Simon & Schuster special sales department at 866-506-1949.

Contents

How to Use This Book

Painless pre-algebra? Impossible, you think. Not really. Math is easy . . . or at least it can be with the help of this book!

Whether you are learning pre-algebra for the first time, or you are trying to remember what you've learned but have forgotten, this book is for you. It provides a clear introduction to pre-algebra that is both fun and instructive. Don't be afraid. Dive in—it's painless!

Painless Icons and Features

This book is designed with several unique features to help make learning pre-algebra easy.

You will see Painless Tips throughout the book. These include helpful tips, hints, and strategies on the surrounding topics.

Caution boxes will help you avoid common pitfalls or mistakes. Be sure to read them carefully.

These boxes translate "math talk" into plain English to make it even easier to understand math.

BRAIN TICKLERS

There are brain ticklers throughout each chapter in the book. These quizzes are designed to make sure you understand what you've just learned and to test your progress as you move forward in the chapter. Complete all the Brain Ticklers and check your answers. If you get any wrong, make sure to go back and review the topics associated with the questions you missed.

PAINLESS STEPS

Complex procedures are divided into a series of painless steps. These steps help you solve problems in a systematic way. Follow the steps carefully, and you'll be able to solve most pre-algebra problems.

EXAMPLES

Most topics include examples with solutions. If you are having trouble, research shows that writing or copying the problem may help you understand it.

ILLUSTRATIONS

Painless Pre-Algebra is full of illustrations to help you better understand pre-algebra topics. You'll find graphs, charts, and more instructive illustrations to help you along the way.

SIDEBARS

These shaded boxes contain extra information that relates to the surrounding topics. Sidebars can include detailed examples or practice tips to help keep pre-algebra interesting and painless.

Chapter 1

Exponents

Exponents

Exponents are a shorthand way of writing repeated multiplication. For example, $2 \cdot 2 \cdot 2 \cdot 2 \cdot 2$ is read as 2 times 2 times 2 times 2 times 2. This multiplication problem can also be written in **exponential form**.

$2 \cdot 2 \cdot 2 \cdot 2 \cdot 2 = 2^5 \rightarrow 2^5$ means the number *2 is a* factor
5 times

If a number is written in exponential form, the exponent tells how many times the base is used as a factor.

5 is the exponent

2 is the base $\rightarrow 2^5$

1+2=3 MATH TALK!

Here are a few ways to say and write exponential expressions:

3^2	4^3	a^4
3 times 3	4 times 4 times 4	a times a times a times a
Three to the second power	Four to the third power	a to the fourth power
3 squared	Four cubed	

To evaluate exponents, just remember that using exponents is just a short way of writing repeated multiplication.

$$3^2 = 3 \cdot 3 = 9 \qquad 3^3 = 3 \cdot 3 \cdot 3 = 27 \qquad 3^4 = 3 \cdot 3 \cdot 3 \cdot 3 = 81$$

CAUTION—Major Mistake Territory!

5^2 means 5 times 5, which equals 25.

5^2 does not mean 5 times 2, which equals ten.

The exponent tells you how many times to multiply the base by itself.

We can also use exponents with negative numbers.

$$(-2)^2 = (-2)(-2) = 4$$
$$(-3)^2 = (-3)(-3) = 9$$
$$(-10)^2 = (-10)(-10) = 100$$

When you square a negative number, the answer is always positive.

$$(-2)^3 = (-2)(-2)(-2) = -8$$
$$(-3)^3 = (-3)(-3)(-3) = -27$$
$$(-4)^3 = (-4)(-4)(-4) = -64$$

When you raise a negative number to the third power, or cube it, the answer is always negative.

PAINLESS TIP

Rules for Exponents of Negative Numbers

A negative number raised to an even power is always a positive number.

A negative number raised to an odd power is always a negative number.

$$(-2)^2 = (-2)(-2) = 4$$
$$(-2)^3 = (-2)(-2)(-2) = -8$$
$$(-2)^4 = (-2)(-2)(-2)(-2) = 16$$
$$(-2)^5 = (-2)(-2)(-2)(-2)(-2) = -32$$
$$(-1)^{100} = 1$$
$$(-1)^{89} = -1$$

BRAIN BUSTER!

When simplifying exponents, watch out for those pesky negative signs. A common mistake is interpreting -3^2 as $(-3)^2$.

-3^2 means the number 3 is squared and the answer is negative.

The negative sign is like multiplying 3^2 by (-1).

It is the same as $(-1)(3)(3) = -9$.

vs.

$(-3)^2$ means the number -3 is squared. So $(-3)(-3) = 9$.

Example 1:

$(-2)^3 = (-2)(-2)(-2) = -8$ vs. $-2^3 = (-1)(2)(2)(2) = -8$

Here, the number (-2) is cubed.   Here, 2 is cubed, and then the answer is negative.

Example 2:

$(-3)^4 = (-3)(-3)(-3)(-3) = 81$ vs. $-3^4 = (-1)(3)(3)(3)(3) = -81$

Another important exponent rule is the **zero power rule**. Any number raised to the zero power equals 1. Be careful what number is being raised to the zero power.

$$3^0 = 1 \qquad 4^0 = 1 \qquad 25^0 = 1 \qquad (-2)^0 = 1 \qquad (-3)^0 = 1$$

BRAIN TICKLERS   Set # 1

Evaluate each of the following.

1. 2^3

2. 4^2

3. 8^2

4. 10^3

5. 7^0

6. $(-3)^3$

7. $(-5)^2$

8. $(-20)^0$

9. -4^2

10. $(-4)^2$

11. -5^2

12. -20^0

(Answers are on page 30.)

Properties of Exponents

PAINLESS TIP

Multiplying Powers with the Same Base

Exponents are a shorthand way of writing repeated multiplication. Look at the following:

$2^2 \cdot 2^2 = 2 \cdot 2 \cdot 2 \cdot 2 = 2^4$
$3^3 \cdot 3^2 = 3 \cdot 3 \cdot 3 \cdot 3 \cdot 3 = 3^5$

When you multiply powers with the same base, keep the base and add the exponents.

Exciting examples:

$5^2 \cdot 5^3 \rightarrow$ Bases are the same. Add the exponents!

5^{2+3}

5^5

$4^{10} \cdot 4^8$	$x^5 \cdot x^{12}$	$a^{15} \cdot a^{-3}$
4^{10+8}	x^{5+12}	$a^{15+(-3)}$
4^{18}	x^{17}	a^{12}

1+2=3 MATH TALK!

Remember what you know about exponents. Write out $6^4 \cdot 6^2$ in standard form.

$6^4 \cdot 6^2$

$(6 \cdot 6 \cdot 6 \cdot 6) \cdot (6 \cdot 6)$

What does $6^4 \cdot 6^2$ equal? 6^6

The shortcut—keep the base, add the exponents.

Let's see what happens when we divide powers with the same base.

$$\frac{4^7}{4^3} = \frac{4 \cdot 4 \cdot 4 \cdot 4 \cdot 4 \cdot 4 \cdot 4}{4 \cdot 4 \cdot 4} = \frac{\cancel{4} \cdot \cancel{4} \cdot \cancel{4} \cdot 4 \cdot 4 \cdot 4 \cdot 4}{\cancel{4} \cdot \cancel{4} \cdot \cancel{4}} = 4 \cdot 4 \cdot 4 \cdot 4 = 4^4$$

PAINLESS TIP

Dividing Powers with the Same Base

To divide powers with the same base, keep the base and subtract the exponents!

More examples:

$\dfrac{5^{12}}{5^4}$	$\dfrac{7^{20}}{7^{14}}$	$\dfrac{x^9}{x^4}$
5^{12-4}	7^{20-14}	x^{9-4}
5^8	7^6	x^5

Let's review!

- To multiply powers with the same base, keep the base and add the exponents.
- To divide powers with the same base, keep the base and subtract the exponents.

1+2=3 MATH TALK!

When multiplying exponents with the same base, add the exponents. When dividing exponents with the same base, subtract the exponents.

Hints to help you:
Times sign × → rotated looks like +
Division sign ÷ → has minus (−) in it

To see what happens when you **raise a power to a power**, we are going to use what we know about writing out exponents and the order of operations.

Example 1:

$(2^3)^4$ — Inside the parentheses, what does 2^3 mean?

$(2 \cdot 2 \cdot 2)^4$ — What does an exponent of 4 mean?

$(2 \cdot 2 \cdot 2) \cdot (2 \cdot 2 \cdot 2) \cdot (2 \cdot 2 \cdot 2) \cdot (2 \cdot 2 \cdot 2) = 2^{12}$ — Write this using one exponent.

Example 2:

$(5^2)^3$
$(5 \cdot 5)^3$
$(5 \cdot 5) \cdot (5 \cdot 5) \cdot (5 \cdot 5)$
5^6

$(5^2)^3$
$\downarrow$
5^6

PAINLESS TIP

Raising a Power to a Power

To raise a power to a power, keep the base and multiply the exponents!

Examples:

$$(4^2)^4 \qquad (6^4)^5 \qquad (x^{12})^2$$

$$4^{2(4)} \qquad 6^{4(5)} \qquad x^{12(2)}$$

$$4^8 \qquad 6^{20} \qquad x^{24}$$

CAUTION—Major Mistake Territory!

You have to keep in mind the difference between multiplying exponents with the same base and raising a power to a power.

$2^3 \cdot 2^5 =$ when you multiply, you **add** the exponents $= 2^8$

$(2^3)^5 =$ power to a power, you **multiply** the exponents $= 2^{15}$

Exponent Rules

Property	Rule	Example
Multiply powers with the same base	Keep the base, add the exponents	$4^5 \cdot 4^6 = 4^{11}$
Divide powers with the same base	Keep the base, subtract the exponents	$\dfrac{7^5}{7^3} = 7^2$
Raise a power to a power	Keep the base, multiply the exponents	$(6^5)^3 = 6^{15}$

BRAIN TICKLERS   Set # 2

Multiply.

1. $5^6 \cdot 5^4$

2. $7^{10} \cdot 7^5$

3. $x^4 \cdot x^{11}$

Divide.

4. $\dfrac{5^9}{5^6}$

5. $\dfrac{8^{15}}{8^4}$

6. $\dfrac{a^{18}}{a^2}$

Simplify.

7. $(3^4)^3$

8. $(4^5)^2$

9. $(x^5)^3$

Multiply. Write the product as one power.

10. $10^5 \cdot 10^5$

11. $m^4 \cdot m^{-2}$

12. $15 \cdot 15^3$

Divide. Write the quotient as one power.

13. $\dfrac{11^8}{11^7}$

14. $\dfrac{14^5}{14^2}$

15. $\dfrac{b^{14}}{b^7}$

(Answers are on page 30.)

Negative Exponents

We just looked at properties of exponents. Remember, an exponent of a number tells you how many times the base is multiplied by itself. So $4^3 = 4 \cdot 4 \cdot 4$. But what about something such as

$$4^{-3}$$

What happens if the exponent is negative?

Let's look for a pattern in the table, using what we already know about exponents.

10^2	10^1	10^0	10^{-1}	10^{-2}	10^{-3}
$10 \cdot 10$	10	1	$\dfrac{1}{10}$	$\dfrac{1}{10 \cdot 10}$	$\dfrac{1}{10 \cdot 10 \cdot 10}$
		(remember, anything to the 0 power = 1)			
100	10	1	$\dfrac{1}{10} = 0.1$	$\dfrac{1}{100} = 0.01$	$\dfrac{1}{1000} = 0.001$

Notice: ÷ 10   ÷ 10   ÷ 10   ÷ 10   ÷ 10

When the exponent is positive, that tells us how many times to multiply the number by itself. So when the exponent is negative, that tells us how many times to divide by the number.

Example:

$$5^{-1} = 1 \div 5 = \frac{1}{5}$$

$$5^{-2} = 1 \div (5 \div 5) = \frac{1}{5^2}$$

$$5^{-3} = 1 \div (5 \div 5 \div 5) = \frac{1}{5^3}$$

> Another way to think about it is $a^{-n} = \dfrac{1}{a^n}$

When any number, except 0, has a negative exponent, take the **reciprocal** of the number with its opposite exponent.

1+2=3 MATH TALK!

The reciprocal of a number is 1 divided by that number.

The reciprocal of 5 is $\dfrac{1}{5}$.

Let's look at a few more.

Negative exponent	Reciprocal with positive exponent	Answer
4^{-2}	$\dfrac{1}{4^2}$	$\dfrac{1}{4} \cdot \dfrac{1}{4} = \dfrac{1}{16}$
10^{-3}	$\dfrac{1}{10^3}$	$\dfrac{1}{10} \cdot \dfrac{1}{10} \cdot \dfrac{1}{10} = \dfrac{1}{1,000}$
2^{-4}	$\dfrac{1}{2^4}$	$\dfrac{1}{2} \cdot \dfrac{1}{2} \cdot \dfrac{1}{2} \cdot \dfrac{1}{2} = \dfrac{1}{16}$
$(-2)^{-3}$	$\dfrac{1}{(-2)^3}$	$\dfrac{1}{-2} \cdot \dfrac{1}{-2} \cdot \dfrac{1}{-2} = \dfrac{1}{-8}$

Write each expression using a positive exponent.

1. $2^{-4} = \dfrac{1}{2^4}$

2. $x^{-3} = \dfrac{1}{x^3}$

3. $7^{-6} = \dfrac{1}{7^6}$

CAUTION—Major Mistake Territory!

When dealing with negative exponents, remember to write the reciprocal of the number and then change the sign of the *exponent*. If the original number is negative, that number stays negative when you take the reciprocal. Evaluate last.

$$(-3)^{-1} = \frac{1}{\left(-3^1\right)} = \frac{1}{-3}$$

$$(-3)^{-2} = \frac{1}{\left(-3\right)^2} = \frac{1}{9}$$

$$(-3)^{-3} = \frac{1}{\left(-3\right)^3} = \frac{1}{-27}$$

Write each fraction using an exponent other than -1.

4. $\dfrac{1}{4^3} = 4^{-3}$ **5.** $\dfrac{1}{m^{10}} = m^{-10}$ **6.** $\dfrac{1}{16} = 2^{-4}$ **7.** $\dfrac{1}{49} = 7^{-2}$

CAUTION—Major Mistake Territory!

When given a problem such as $\dfrac{1}{81}$ and asked to rewrite it using an exponent other than -1, you cannot use 81^{-1}. You have to think of a number that is raised to a power to get the number you are looking for. So $\dfrac{1}{81} = 9^{-2}$.

BRAIN TICKLERS   Set # 3

Now you try! Write using a positive exponent and then evaluate.

1. 8^{-2}

2. $(-5)^{-2}$

3. -5^2

4. w^{-12}

5. 3^{-4}

6. 10^{-2}

7. $(-4)^{-2}$

8. $4a^{-3}$

(Answers are on page 30.)

Operations with Negative Exponents

Use your laws of exponents and your rule for negative exponents to try these problems. Let's quickly review our rules before we start.

Properties	Example	Process	Answer
Product rule			
Keep base	$6^2 \cdot 6^5$	6^{2+5}	6^7
Add exponents	$(x^{10})(x^5)$	x^{10+5}	x^{15}
Quotient rule			
Keep base	$\dfrac{12^8}{12^6}$	12^{8-6}	12^2
Subtract exponents	$\dfrac{m^{17}}{m^8}$	m^{17-8}	m^9
Power rule			
Keep base	$(8^3)^6$	$8^{3 \cdot 6}$	8^{18}
Multiply exponents	$(n^{12})^5$	$n^{12 \cdot 5}$	n^{60}

Properties	Example	Process	Answer
Negative exponent			
Take reciprocal (flip)	2^{-3}	$\dfrac{1}{2^3}$	$\dfrac{1}{8}$
Make exponent positive	x^{-5}	$\dfrac{1}{x^5}$	$\dfrac{1}{x^5}$
Zero exponent			
Anything to the zero power (except 0) is 1	5^0	$5^0 = 1$	1
	$(-3)^0$	$(-3)^0 = 1$	1

Note! If bases are not the same, rules do not apply!

For example, $\dfrac{x^{17}}{y^8}$ cannot be simplified.

Examples:

Simplify.

a) $5^5 \cdot 5^{-2} = 5^{5 + (-2)} = 5^3$

b) $\dfrac{12^{10}}{12^6} = 12^{10 - 6} = 12^4$

c) $\dfrac{12^8}{12^{-6}} = 12^{8 - (-6)} = 12^{14}$

d) $7^{-12} \cdot 7^{-4} = 7^{-12 + (-4)} = 7^{-16} = \dfrac{1}{7^{16}}$

e) $(6^{-10})(6^5) = 6^{-10 + 5} = 6^{-5} = \dfrac{1}{6^5}$

f) $(5^2)^{-3} = 5^{2(-3)} = 5^{-6} = \dfrac{1}{5^6}$

g) $(7^{-3})^{-5} = 7^{-3 \cdot -5} = 7^{15}$

h) $\dfrac{z^{-4}}{z^{-3}} = z^{-4 - (-3)} = z^{-4 + 3} = z^{-1} = \dfrac{1}{z}$

Answer these questions based on the previous examples. Answers are listed below.

1. Which examples used the product rule?

2. Which examples used the quotient rule?

3. Which examples used the power rule?

4. Which examples used an exponent rule and the negative exponent rule?

Answers:

1. a, e, d
2. b, c, h
3. f, g
4. d, e, f, h

Notice there is a lot of math for each example. You have to remember integer rules, exponent rules, and the negative exponent rule. Take one step at a time, and you will get it!

BRAIN TICKLERS   Set # 4

Express using positive exponents.

1. $2^{-6} \cdot 2^3$

2. $x^{-5} \cdot x^7$

3. $\dfrac{10^7}{10^{12}}$

4. $\dfrac{3^7}{3^7}$

5. $(b^9)^{-3}$

6. $(m^{-3})^{-4}$

7. $\dfrac{4^{-8}}{4^7}$

8. $(7^{-8})(7^{-3})$

9. $(m^{-3})^{-4}$

10. Write $\dfrac{1}{81}$ using a negative exponent other than -1.

(Answers are on page 31.)

Scientific Notation

Scientific notation is a shorthand way of expressing really large numbers or very small numbers. Understanding exponents makes understanding scientific notation easy! There are two parts to keep in mind when writing a number in scientific notation.

$$2.5 \times 10^5$$

The first number has to be greater than or equal to 1 and less than 10.

The second number is 10 to a power. The power will show how many places to move the decimal point.

Numbers written in scientific notation	Numbers not in scientific notation and reason		
3.2×10^3	0.1×10^3	$\rightarrow$	First number is less than 1
4×10^{-4}	12×10^6	$\rightarrow$	First number larger than 10
9.65×10^{23}	3×1^2	$\rightarrow$	Second number has to be times 10 to a power

To translate a number **from scientific notation to standard form**, you need to look at the exponent on the 10. If the exponent is positive, move the decimal to the right. If the exponent is negative, move the decimal to the left. You move the decimal point as many times as the exponent indicates.

PAINLESS TIP

Decimal Point Direction

Positive exponent moves decimal to the right $\rightarrow$.

Negative exponent moves decimal to the left $\leftarrow$.

Example 1:

Write 3.25×10^5 in standard form.

10^5 is positive, so move the decimal five places to the right.

3.25×10^5

3.25000

$\rightarrow$

325,000

Example 2:

Write 4.5×10^{-3} in standard form.

10^{-3} is negative, so move the decimal three places to the left.

4.5×10^{-3}

004.5

$\leftarrow$

0.0045

Let's summarize.

- Write the digits with the decimal point placed after the first digit.
- Write $\times$ 10 to a power. The power shows how many places to move the decimal point.

Scientific notation	Standard form	Scientific notation	Standard form
2.3×10^2	230	4.5×10^{-1}	0.45
2.3×10^3	2,300	4.5×10^{-2}	0.045
2.3×10^4	23,000	4.5×10^{-3}	0.0045

Now let's go the other way. Let's translate a number **from standard form to scientific notation.**

Example 3:

Take the number 25,000. Put the decimal after the first digit, keep all other different digits, but drop the repeated zeros.

$\boxed{2.5}000$

2.5 becomes the coefficient, which is the first part of the scientific notation form. → $2.5 \times 10^?$

Find the exponent by counting the number of places from the decimal to the end of the number. Since we will be moving right, the exponent will be positive.

In 25,000, there are four places. Therefore, we write 25,000 as 2.5×10^4.

Example 4:

Let's try a decimal. Take the number 0.0032. Put the decimal after the first digit.

$000\boxed{3.2}$

3.2 becomes the coefficient → $3.2 \times 10^?$. Find the exponent by counting the number of places from the new decimal to the decimal in the original number. Since we will be moving left, the exponent will be negative.

In 003.2, there are three places. Therefore, we write 0.0032 as 3.2×10^{-3}.

Let's Review Scientific Notation!

Place the decimal point after the **first digit**	Express 5,245 in scientific notation 5,245 → 5.245	Express 0.00045 in scientific notation 0.00045 → 4.5
Write the new number × 10	$5.245 \times 10^?$	$4.5 \times 10^?$
To find the power of 10, count the number of places from the decimal to get the original number:	5.245 → 5,245	0004.5 → 0.00045
	Decimal moved three places right	Decimal moved four places left

(large number → positive exponent)
(decimal number → negative exponent)

Scientific notation	$\mathbf{5.245 \times 10^3}$	$\mathbf{4.5 \times 10^{-4}}$

BRAIN TICKLERS   Set # 5

Write each number in standard form.

1. 2.4×10^7

2. 3.65×10^{10}

3. 7.102×10^4

4. 5.2×10^{-3}

5. 2.9×10^{-1}

6. 4.56×10^{-6}

Write each number in scientific notation.

7. 234,000

8. 1,123

9. 307,000,000

10. 0.0005

11. 0.032

12. 0.00000456

(Answers are on page 31.)

Multiplication and Division with Scientific Notation

You can use the product of powers and the quotient of powers properties to multiply and divide numbers written in scientific notation. Using your laws of exponents makes operations with big numbers easy!

Examples:

1. Evaluate $(2.63 \times 10^4)(1.2 \times 10^{-3})$

The math	In words
$(2.63 \times 10^4)(1.2 \times 10^{-3})$	Write original problem
$(2.63 \times 1.2)(10^4 \times 10^{-3})$	Group decimals, group exponents
3.156×10^1	(multiply decimals) × (keep base, add exponents)
3.156×10^1	Write final answer in scientific notation (It is already, so you are done!)

Easy right? Look at this one!

2. Evaluate $(8.4 \times 10^2)(2.5 \times 10^6)$

The math	In words
$(8.4 \times 10^2)(2.5 \times 10^6)$	Write original problem
$(8.4 \times 2.5)(10^2 \times 10^6)$	Group decimals, group exponents
21×10^8	(multiply decimals) × (keep base, add exponents)
2.1×10^9	Write final answer in scientific notation

Remember, to be in scientific notation, the number must be greater than 1 and less than 10. So 21 needs to be corrected, and you will have to adjust the exponent. To help you, write 21×10^8 in standard form.

$$21 \times 10^8 = 2,100,000,000$$

Put decimal after the 2, so it is 2.100000000.

And rewrite in scientific notation, 2.1×10^9.

3. Evaluate $(7.2 \times 10^3)(1.6 \times 10^4)$. Express in scientific notation.

The math	In words
$(7.2 \times 10^3)(1.6 \times 10^4)$	Write original problem
$(7.2 \times 1.6)(10^3 \times 10^4)$	Group decimals, group exponents
11.52×10^7	(multiply decimals) $\times$ (keep base, add exponents)
1.152×10^8	Write final answer in scientific notation

Not in scientific notation. Shortcut = move decimal back one spot, add one to exponent.

How could you check yourself?

Don't use scientific notation! Write it out, multiply, and write your answer in scientific notation!

$$(7.2 \times 10^3)(1.6 \times 10^4) = (7,200)(16,000) =$$
$$115,200,000 = 1.152 \times 10^8. \text{ It works!}$$

If you can multiply in scientific notation, then it is super easy to divide in scientific notation. Check this out! Remember your law of exponent rules for division!

4. Evaluate: $\dfrac{8.64 \times 10^6}{4.32 \times 10^2}$

The math	In words
$\dfrac{8.64 \times 10^6}{4.32 \times 10^2}$	Write original problem
$\left(\dfrac{8.64}{4.32}\right)\left(\dfrac{10^6}{10^2}\right)$	Group decimals, group exponents
2×10^4	(divide decimals) $\times$ (keep base, subtract exponents)
2×10^4	Write final answer in scientific notation (It is already, so you are done!)

How could you check your answer?

Write the numbers in standard form.

$$\frac{8,640,000}{432} = 8,640,000 \div 432 = 20,000$$

which equals 2×10^4. It checks!

5. Evaluate: $\dfrac{2.016 \times 10^7}{8.4 \times 10^3}$

The math	In words
$\dfrac{2.016 \times 10^7}{8.4 \times 10^3}$	Write original problem
$\left(\dfrac{2.016}{8.4}\right)\left(\dfrac{10^7}{10^3}\right)$	Group decimals, group exponents
0.24×10^4	(divide decimals) × (keep base, subtract exponents)
2.4×10^3	Write final answer in scientific notation

Write 0.24×10^4 in standard form.

$$0.24 \times 10^4 = 2,400$$

Put decimal after the 2, so it is 2.400.

Rewrite in scientific notation, 2.4×10^3.

1+2=3 MATH TALK!

When writing numbers so they are in correct scientific notation:

If you move the decimal to the left, add to the exponent.

   35×10^7 becomes 3.5×10^8

If you move the decimal to the right, subtract from the exponent.

   0.345×10^{12} becomes 3.45×10^{11}

BRAIN TICKLERS   Set # 6

Evaluate each expression. Express the result in scientific notation. (Be careful when adding and subtracting exponents!)

1. $(3.2 \times 10^4)(1.4 \times 10^2)$

2. $(6.1 \times 10^5)(8.2 \times 10^4)$

3. $(5.2 \times 10^{-5})(7.4 \times 10^3)$

4. $(2 \times 10^{-2})(5.5 \times 10^{-5})$

5. $\dfrac{4.55 \times 10^7}{1.3 \times 10^3}$

6. $\dfrac{8.05 \times 10^4}{2.3 \times 10^{-3}}$

7. $\dfrac{1.175 \times 10^{-3}}{1.25 \times 10^{-8}}$

8. $\dfrac{5 \times 10^7}{4 \times 10^{10}}$

(Answers are on page 31.)

Adding and Subtracting in Scientific Notation

Do you remember how to add 345 and 52? You stacked the numbers so you could add correct place values!

$$\begin{array}{r} 345 \\ +\ \ 52 \\ \hline 397 \end{array}$$

This concept still applies when adding and subtracting numbers in scientific notation. But to get the place values to line up, we have to pay special attention to the exponents.

To add and subtract in scientific notation:

1. Adjust the powers of 10 in both numbers so they have the same exponent. (Hint: Make smaller exponent into larger exponent; it's easier!)

2. Add or subtract the decimals.

3. Write the number in scientific notation.

Examples:

1. $(2 \times 10^3) + (3.6 \times 10^4)$

This is the smaller number, so we need to change this to have an exponent of 4. To add an exponent, move decimal BACK number of spaces needed. In this case, move back once—this adds one to the exponent.

2×10^3 will become 0.2×10^4.

The math	In words
$(2 \times 10^3) + (3.6 \times 10^4)$	Write original problem
$(0.2 \times 10^4) + (3.6 \times 10^4)$	Change 2×10^3
$(0.2 + 3.6) \times 10^4$	Group decimals, keep $\times 10^4$
3.8×10^4	Add decimals, keep $\times 10^4$
3.8×10^4	Make sure final answer is in scientific notation

Check yourself by writing the numbers in standard form and adding them.

$2 \times 10^3 = 2,000$ and $3.6 \times 10^4 = 36,000$.
So $36,000 + 2,000 = 38,000$, which is 3.8×10^4.

2. $(7 \times 10^5) - (5.2 \times 10^3)$

The math	In words
$(7 \times 10^5) - (5.2 \times 10^3)$	Write original problem
$(7 \times 10^5) - (0.052 \times 10^5)$	Change 5.2×10^3 (need 3 to be 5, so move decimal back 2 spots, which adds 2 to the exponent)
$(7 - 0.052) \times 10^5$	Group decimals, keep $\times 10^5$
6.948×10^5	Subtract decimals, keep $\times 10^5$
6.948×10^5	Make sure final answer is in scientific notation

1+2=3 MATH TALK!

When adding and subtracting in scientific notation, exponents must MATCH.

To get **smaller exponent to become a larger exponent**, move **LEFT** the number of spaces you need.

$2.4 \times 10^7 + 5.6 \times 10^5$ 5.6×10^5 needs an exponent of 7, so move decimal LEFT two places and add 2 to the exponent.

5.6×10^5 becomes 0.056×10^7.

3. $(6.3 \times 10^5) + 2{,}700{,}000$

The math	In words
$(6.3 \times 10^5) + 2{,}700{,}000$	Write original problem
$(6.3 \times 10^5) + (2.7 \times 10^6)$	2,700,000 needs to be in scientific notation
$(0.63 \times 10^6) + (2.7 \times 10^6)$	Change 6.3×10^5 to have exponent of 6 (need 5 to be 6, so move decimal back 1 spot, which adds 1 to the exponent)
$(0.63 + 2.7) \times 10^6$	Group decimals, keep $\times\, 10^6$
3.33×10^6	Add decimals, keep $\times\, 10^6$
3.33×10^6	Make sure final answer is in scientific notation

Cool tip! By changing the smaller exponent to the larger, your final answer will always be in scientific notation!

BRAIN TICKLERS   Set # 7

Add or subtract each of the following. Express your answer in scientific notation and standard form.

1. $(2.7 \times 10^3) + (3.4 \times 10^2)$
2. $(9.2 \times 10^3) - (9.6 \times 10^2)$
3. $(8.4 \times 10^5) - (7.9 \times 10^3)$
4. $593{,}000 + (7.89 \times 10^6)$

(Answers are on page 32.)

Applications of Operations with Scientific Notation

Understanding the skill behind multiplying, dividing, adding, and subtracting numbers written in scientific notation will help you solve real-life problems involving large and small numbers.

PAINLESS TIP

When tackling these problems, ask yourself:

- Are all the numbers expressed in the same format? If not, make them all the same (all scientific notation or all standard form).
- What key words help me decide if I am adding, subtracting, multiplying, or dividing?
- Is my final answer in scientific notation?
- Does my answer make sense?

Example 1:

Did you know that 150,000,000,000 spam emails are sent worldwide every day? How many emails would be sent in 365 days (1 year)? Express your answer in scientific notation.

Solution:
Write both numbers in scientific notation:

$150,000,000,000 = 1.5 \times 10^{11}$

$365 = 3.65 \times 10^2$

Multiply since it is a certain number, every day, for 365 days:

$(1.5 \times 10^{11})(3.65 \times 10^2)$

$(1.5 \times 3.65)(10^{11} \times 10^2)$

5.475×10^{13}

Does 5.475×10^{13} make sense? Is that a lot of email? You bet!

Example 2:

In 2015, the world population was about 7.2 billion. The population of the United States in 2015 was about 320 million. About how many times larger was the world population than the population of the United States?

Solution:

7.2 billion = 7,200,000,000 = 7.2×10^9

320 million = 320,000,000 = 3.2×10^8

How many times larger? Division problem.

$$\frac{7.2 \times 10^9}{3.2 \times 10^8} = \left(\frac{7.2}{3.2}\right)\left(\frac{10^9}{10^8}\right) = 2.25 \times 10$$

What does 2.25×10 mean to answer the question? How many times larger?

$2.25 \times 10 = 22.5$, which means the world population was about 22.5 times larger than the United States population.

How could you check yourself? Does your answer make sense?

$$\frac{7,200,000,000}{320,000,000} = 22.5$$ And what does

$320,000,000 \times 22.5 = ?$

$320,000,000 \times 22.5 = 7,200,000,000$

Example 3:

The attendance records for two Major League Baseball teams are shown below.

Which team has the greater attendance and by how much?

Team	Attendance
Miami Marlins	6.78×10^5
St. Louis Cardinals	1.97×10^6

Solution:

The St. Louis Cardinals have the greater attendance. You can tell this because it has the larger exponent, which makes it a larger number.

By how much? This is a subtracting problem.

Solution using scientific notation	Solution using standard form
$(1.97 \times 10^6) - (6.78 \times 10^5)$	$1,970,000 - 678,000$
$(1.97 \times 10^6) - (0.678 \times 10^6)$	$1,292,000$
$(1.97 - 0.678) \times 10^6$	1.292×10^6
1.292×10^6	

To answer the question, it makes more sense to use the standard form number. The St. Louis Cardinals have 1,292,000 more people in attendance than the Miami Marlins.

Example 4:

A computer hard drive can hold 8×10^7 bytes of information. A memory stick can hold 2×10^6 bytes of information. How much memory would you have altogether?

Solution:
This is an addition problem since they are asking how much altogether.

$(8 \times 10^7) + (2 \times 10^6)$

$(8 \times 10^7) + (0.2 \times 10^7)$

$(8 + 0.2) \times 10^7$

8.2×10^7

The computer and memory card will hold 8.2×10^7 bytes of data or 82,000,000 bytes.

BRAIN TICKLERS   Set # 8

1. Jupiter has a diameter of about 8.8×10^4 miles. Mercury has a diameter of about 3×10^3 miles. What is the difference in their diameters? Express your answer in scientific notation.

2. Bernie loves to make bead bracelets. She bought a box of 5,000 beads. Bernie used 3.78×10^3 beads last month and used 2.64×10^2 beads this month. How many beads has Bernie used in the past two months? How many beads will be left over after these two months?

3. Central Park in New York City is in the shape of a rectangle. The length of the park is about 1.34×10^4 feet, and the width is approximately 2.55×10^2 feet. How many more feet is the length compared to the width? What is the area of the park?

4. SongTunes announced that 6×10^{11} songs were downloaded by users just this year. If 5×10^8 users are signed up for SongTunes, what is the average number of downloads per user? If last year's average was 980 songs per user, is this year's average more or less?

(Answers are on page 32.)

BRAIN TICKLERS—THE ANSWERS
Set # 1, page 4

1. 8

2. 16

3. 64

4. 1,000

5. 1

6. -27

7. 25

8. 1

9. -16

10. 16

11. -25

12. -1

Set # 2, page 8

1. 5^{10}

2. 7^{15}

3. x^{15}

4. 5^3

5. 8^{11}

6. a^{16}

7. 3^{12}

8. 4^{10}

9. x^{15}

10. 10^{10}

11. m^2

12. 15^4

13. 11

14. 14^3

15. b^7

Set # 3, page 12

1. $\dfrac{1}{8^2} = \dfrac{1}{64}$

2. $\dfrac{1}{(-5)^2} = \dfrac{1}{25}$

3. -25

4. $\dfrac{1}{w^{12}}$

5. $\dfrac{1}{3^4} = \dfrac{1}{81}$

6. $\dfrac{1}{10^2} = \dfrac{1}{100}$

7. $\dfrac{1}{(-4)^2} = \dfrac{1}{16}$

8. $\dfrac{4}{a^3}$

Set # 4, page 14

1. $\dfrac{1}{2^3}$

2. x^2

3. $\dfrac{1}{10^5}$

4. 1

5. $\dfrac{1}{b^{27}}$

6. m^{12}

7. $\dfrac{1}{4^{15}}$

8. $\dfrac{1}{7^{11}}$

9. m^{12}

10. 9^{-2}

Set # 5, page 18

1. 24,000,000

2. 36,500,000,000

3. 71,020

4. 0.0052

5. 0.29

6. 0.00000456

7. 2.34×10^5

8. 1.123×10^3

9. 3.07×10^8

10. 5×10^{-4}

11. 3.2×10^{-2}

12. 4.56×10^{-6}

Set # 6, page 22

1. 4.48×10^6

2. 5.002×10^{10}

3. 3.848×10^{-1}

4. 1.1×10^{-6}

5. 3.5×10^4

6. 3.5×10^7

7. 9.4×10^4

8. 1.25×10^{-3}

Set # 7, page 25

1. 3.040×10^3 or 3,040

2. 8.24×10^3 or 8,240

3. 8.321×10^5 or 832,100

4. 8.483×10^6 or 8,483,000

Set # 8, page 29

1. 8.5×10^4 miles

2. 4.044×10^3 or 4,044 beads all together. 956 (9.56×10^2) beads left over.

3. 13,145 ft (1.3145×10^4). The area is 3.417×10^6 sq ft (3,417,000).

4. 1.2×10^3 (1,200 songs). This year's average is more.

Chapter 2

Solving Equations

Two-Step Equations

Sometimes more than one operation is needed to solve equations. Let's review how to solve two-step equations.

Step 1: Undo addition or subtraction.

Step 2: Undo multiplication or division.

Exciting examples:

1. Solve: $2x + 3 = 15$

Step 1: Undo addition or subtraction. Since the problem says $+3$, subtract 3 from both sides of the equation.

$$\begin{array}{r} 2x + 3 = 15 \\ \underline{-3\ -3} \\ 2x\quad = 12 \end{array}$$

Step 2: Undo multiplication or division. Since the problem says "2 times x equals 12," divide both sides by 2.

$$\frac{2x}{2} = \frac{12}{2}$$
$$x = 6$$

Let's try another one!

2. Solve: $\dfrac{x}{3} - 5 = 4$

Step 1: Undo addition or subtraction. Since the problem says -5, add 5 to both sides.

$$\begin{array}{r} \dfrac{x}{3} - 5 = 4 \\ \underline{+5 \ +5} \\ \dfrac{x}{3} \quad = 9 \end{array}$$

Step 2: Undo multiplication or division. The problem reads "x divided by 3 equals 9." To undo division, multiply by 3.

$$3\left(\dfrac{x}{3}\right) = 9(3)$$
$$x = 27$$

To check, just substitute the answer you got for the variable in the original equation. If both sides of the equation are equal, your answer is correct!

Let's check our last problem. We solved $\dfrac{x}{3} - 5 = 4$ and found $x = 27$.

Step 1: Rewrite the problem.

$$\dfrac{x}{3} - 5 = 4$$

Step 2: Plug in the answer, 27, for x.

$$\dfrac{27}{3} - 5 = 4$$

Step 3: Simplify each side using the order of operations.

$$9 - 5 = 4$$
$$4 = 4 \text{ Check!}$$

Your equation works; therefore, 27 is the correct answer.

3. Sophie received a $95 gift card to her favorite clothing store in the mall for her birthday. She wants to purchase some leggings that cost $15 each. She also wants to buy one bottle of perfume for $20. Write an equation that could represent the number of leggings Sophie could buy. Then solve the equation.

Solution:
Since the leggings cost $15 EACH and we want to know how many leggings she can buy, let $x =$ number of leggings.

Equation:

$$15x + 20 = 95$$
$$\underline{-20 \quad -20}$$
$$\frac{15x}{15} = \frac{75}{15}$$
$$x = 5$$

Sophie can buy 5 pairs of leggings.

4. On vacation, Sam wants to rent an electric scooter for one day. Scoot-A-Long charges $28 plus $0.25 per mile to rent a scooter. If Sam only has $40, how many miles can he ride his scooter?

Solution:

Define a variable: Let $m =$ number of miles Sam can ride the scooter

Write an equation: $28 + 0.25m = 40$

Solve:

$$28 + 0.25m = 40$$
$$\underline{-28 \qquad\qquad -28}$$
$$\frac{0.25m}{0.25} = \frac{12}{0.25}$$
$$m = 48$$

Check:

$$28 + 0.25(48) = 40$$
$$28 + 12 = 40$$
$$40 = 40 \text{ Check!}$$

Sam can ride 48 miles on his scooter.

BRAIN TICKLERS   Set # 9

Solve these two-step equations.

1. $6x - 3 = 15$

2. $4x + 1 = 9$

3. $-3x + 3 = 15$

4. $\dfrac{x}{4} + 3 = 10$

5. $\dfrac{x}{2} - 4 = 8$

6. Cayden wants to buy a new bike that costs $191. This is $25 less than four times what he saved last month for the bike. How much money did Cayden save last month to buy this bike?

7. Sixteen less than one-fourth of a number is equal to -20. Find the number.

8. Kaleb has $45 to spend at the carnival. If it costs $2 per ride and Kaleb spends $15 on food, how many rides can Kaleb go on?

(Answers are on page 55.)

Combining Like Terms

Before we can solve more equations, we need to take a look at like terms. **Terms** in an expression are separated by plus or minus signs.

$$8x + 4y - 3x + 10y$$

In the above example, there are four terms: $8x$, $4y$, $-3x$, and $10y$. When identifying terms, you have to remember that the sign in front of the number goes with the term. Let's look at a few more.

Expression	# of terms	List the terms
$k + 3g - 4k + 10$	4	$k, 3g, -4k, 10$
$7a + 4a - 3b - 4b + 7$	5	$7a, 4a, -3b, -4b, 7$

Like terms, such as $8x$ and $5x$, can be grouped together because they have the same variable, raised to the same exponent. Let's look at a few examples.

$3x$ and $4y$	NOT like terms	The variables (letters) are different
$3x$ and 7	NOT like terms	The 7 does not have a matching x variable
$4x$ and $3x^2$	NOT like terms	The exponent on the second term, $3x^2$, does not match the exponent on the first term
$3x$ and $4x$	LIKE TERMS!	Both terms have the variable x raised to the first power
$5x^2$ and $-7x^2$	LIKE TERMS!	Both terms have the variable x raised to the second power

Now that we can identify like terms, we can use this to help make simplifying expressions a lot easier. When trying to combine like terms, a good strategy is to use "Circle, Box, Underline." Circle terms that are alike, box other terms that are alike, and underline other terms that are alike. It's painless, you'll see.

Combine like terms

$5x + 3x$

$(5x) + (3x)$　　　　　　　Circle $5x$ and $3x$ since they are like terms.

$8x$　　　　　　　　　　　Combine the coefficients: $5 + 3 = 8$.

$5a + 7 - a + 12$

$(5a) \boxed{+7} (-a) \boxed{+12}$　　Box and circle like terms.

$4a + 19$　　　　　　　　Combine coefficients: $5a - a$, and $7 + 12$.

CAUTION–Major Mistake Territory!

The coefficient of a variable by itself, such as k, is 1, because $1k = k$.

$7a + 3a + 4b + 5 + 6b$

$\boxed{7a} + \boxed{3a} \boxed{+4b} \pm 5 \boxed{+6b}$ Circle, box, and underline like
terms.

$10a + 10b + 5$ Combine coefficients:
$7a + 3a, 4b + 6b,$ and 5.

$-9m + 13n - 8m - 6n - 12z$

$\boxed{-9m} \boxed{+13n} \boxed{-8m} \boxed{-6n} \underline{-12z}$ Circle, box, and underline like
terms.

$-17m + 7n - 12z$ Combine coefficients:
$-9m - 8m, 13n - 6n, -12z$.

$18x + 3x^2 - 15x - 2x^2 + 2x + 3$

$\boxed{18x} + \boxed{3x^2} \boxed{-15x} \boxed{-2x^2} \boxed{+2x} \underline{+3}$ Circle, box, and underline like
terms.

$5x + x^2 + 3$ Combine coefficients:
$18x - 15x + 2x, 3x^2 - 2x^2, 3$.

 BRAIN TICKLERS   Set # 10

Combine like terms.

1. $5x + 12x$

2. $14y - 8 - 10y$

3. $3a + 6 - 2a + 12$

4. $16c + 14d + 10 - 8c - 6d - 2$

5. $7d - d + 5e + 23$

(Answers are on page 55.)

The Distributive Property

Another important step to simplifying expressions and equations is
to use the **distributive property**.

What happens when a teacher distributes papers to the class? The teacher will hand out a piece of paper to every student in the class. Well, how does this relate to math? Let's look at an example.

$$2(6 + 3)$$

There are actually two ways to simplify this. The first is to use the order of operations. When using the order of operations, you simplify in the parentheses first. $2(6 + 3) = 2(9) = 18$.

The second way to simplify $2(6 + 3)$ is to use the distributive property. The distributive property allows you to multiply first! The 2 is multiplied (distributed) to each number inside the parentheses, which looks like this:

$$2(6 + 3) = 2(6) + 2(3)$$
$$= 12 + 6$$
$$= 18$$

Remember, the 2 must be distributed (handed out by multiplying) to <u>each</u> term inside the parentheses.

PAINLESS TIP

The Distributive Property

To multiply a number by a sum, multiply each number in the sum by the number in front of the parentheses.

$a(b + c) = ab + ac$
$a(b - c) = ab - ac$

Let's try a few! Simplify using the distributive property.

$3(N + 2)$	$4(k - 3)$	$3(3x + 2y)$
$3(N) + 3(2)$	$4(k) - 4(3)$	$3(3x) + 3(2y)$
$3N + 6$	$4k - 12$	$9x + 6y$

Now that you know how to distribute and combine like terms, we can start to do problems such as this!

$3(y + 4) + 2y$
$3(y) + 3(4) + 2y$ Distribute.
$3y + 12 + 2y$ Multiply.
$5y + 12$ Combine like terms.

$5(y - 2) - 10$
$5(y) - 5(2) - 10$ Distribute.
$5y - 10 - 10$ Multiply.
$5y - 20$ Combine like terms.

1+2=3 MATH TALK!

When simplifying expressions, first you clear the parentheses by using the distributive property, and then you combine like terms.

$3(x + y) + 12x$
$3(x) + 3(y) + 12x$
$3x + 3y + 12x$
$15x + 3y$

BRAIN TICKLERS   Set # 11

Simplify.

1. $5(y + 2)$

2. $3(2y + 8)$

3. $4(x - 7)$

4. $3(5x - 1)$

5. $2(3x - 5) + 10$

6. $3(x + 5) + 2x$

(Answers are on page 55.)

Multistep Equations

You are a pro at solving two-step equations, so multistep equations will be easy! Sometimes, before you can undo addition or subtraction, you might have to simplify the equation. So that will become a new step. These steps will work for solving any equation; just skip step 1 if it is not a multistep equation!

Step 1: Simplify the equation.
> → Do you have to distribute?
> → Do you have to combine like terms?

Step 2: Undo addition or subtraction.

Step 3: Undo multiplication or division.

Examples:

Solve: $2x + 5x - 4 = 17$

Step 1: Simplify the equation by combining like terms.
$2x + 5x - 4 = 17$
$7x - 4 = 17$

Step 2: Undo addition or subtraction
$$7x - 4 = 17$$
$$\underline{+4 \ +4}$$
$$7x \quad = 21$$

Step 3: Undo multiplication or division
$$\frac{7x}{7} = \frac{21}{7}$$
$$x = 3$$

Painless! Let's try another!

Solve: $2(x + 4) = 32$

Step 1: Simplify the equation by distributing.
$2(x + 4) = 32$
$2x + 8 = 32$

Step 2: Undo addition or subtraction.

$$2x + 8 = 32$$
$$\underline{\quad -8 \quad -8\quad}$$
$$2x \quad\;\; = 24$$

Step 3: Undo multiplication or division.

$$\frac{2x}{2} = \frac{24}{2}$$
$$x = 12$$

Here is a third example.

Solve: $4(x + 3) + 2x = 48$

Step 1: Simplify by distributing and then combining like terms.

$$4(x + 3) + 2x = 48$$
$$4x + 12 + 2x = 48$$
$$6x + 12 = 48$$

Step 2: Undo addition or subtraction.

$$6x + 12 = 48$$
$$\underline{\quad -12\; -12\quad}$$
$$6x \quad\;\; = 36$$

Step 3: Undo multiplication or division.

$$\frac{6x}{6} = \frac{36}{6}$$
$$x = 6$$

Let's check our third example.

Step 1: Rewrite the original problem.

$$4(x + 3) + 2x = 48$$

Step 2: Substitute 6 for x.

$$4(6 + 3) + 2(6) = 48$$

Step 3: Simplify each side using the order of operations.

$$4(9) + 12 = 48$$
$$36 + 12 = 48$$
$$48 = 48 \text{ Check!}$$

More examples:

Solve: $4y - 2(y - 5) = -2$

Step 1: Simplify by distributing the -2. Watch your signs! Then combine like terms.

$$4y - 2(y - 5) = -2$$
$$4y - 2y + 10 = -2$$
$$2y + 10 = -2$$

Step 2: Finish the second step—looking familiar, right?!

$$2y + 10 = -2$$
$$\underline{-10 \quad -10}$$
$$\frac{2y}{2} = \frac{-12}{2}$$
$$y = -6$$

Solve: $4 - 3(x + 2) - 2(x - 4) = 1$

Steps: $4 - 3x - 6 - 2x + 8 = 1$ Distribute!

$$-5x + 6 = 1$$ Combine like terms!
$$\underline{-6 \quad -6}$$
$$\frac{-5x}{-5} = \frac{-5}{-5}$$
$$x = 1$$

Apply what you know to set up and solve this problem using an equation!

Remember, when solving problems, use strategies to help you:

- Draw a picture
- Label
- Define your unknown variables

- Read and reread the question
- Ask yourself if the answer makes sense

The length of a rectangle is 5 inches more than twice its width. If the perimeter is 100 inches, find the length of the rectangle. What is the area of the rectangle?

What do we know?

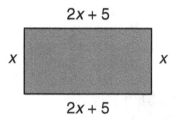

2x + 5

x x

2x + 5

Perimeter means the distance around the outside, so add all the sides! And solve for x.

$$2x + 5 + x + 2x + 5 + x = 100$$
$$6x + 10 = 100$$
$$\underline{-10 \quad -10}$$
$$\frac{6x}{6} = \frac{90}{6}$$
$$x = 15$$

Since $x = 15$, we know the width is 15. REREAD the question. It asks for the <u>length</u>. The length is represented by $2x + 5$, so the length is $2(15) + 5$, which equals 35. The area of a rectangle is length × width, so the area is $(15)(35) = 525$ square units.

Does your answer make sense?

Width = 15, length = 35

Perimeter = 15(2) + 35(2)

30 + 70 = 100! It works!

BRAIN TICKLERS   Set # 12

Solve each of the following equations and check.

1. $3(x - 4) = 24$

2. $4x + 3x + 7x + 5 = 33$

3. $2(x + 1) - 10 = 22$

4. $-3x + 5(x + 1) = 11$

5. $\frac{1}{2}(x + 8) - 15 = -3$

6. $3(x - 2) + 2(x + 1) = -14$

7. The sides of a hexagon are increased by 2 units. If the perimeter of the new hexagon is 72 centimeters, find the length of one side of the original hexagon.

(Answers are on page 55.)

Variables on Both Sides

Some equations have variables on both sides of the equals sign. Solving an equation with variables on both sides is similar to solving an equation with a variable on only one side. We are used to solving problems like this:

$$2x - 5 = 11$$

The first step was to undo addition or subtraction. Why? You are combining like terms to get all of the numbers on one side of the equation.

$$
\begin{aligned}
2x - 5 &= 11 \\
+5 \quad &+5 \\
\hline
\frac{2x}{2} &= \frac{16}{2} \\
x &= 8
\end{aligned}
$$

For a problem with variables on both sides, you still have to undo addition or subtraction. However, now you also have to move the variables!

Examples:

Solve: $2x + 12 = 6x$

In this problem, we have to get all of the x variables on one side of the equation.

Step 1: $2x + 12 = 6x$

Subtract $2x$ from both sides of the equation.

$$\begin{array}{r} 2x + 12 = 6x \\ -2x \quad\quad -2x \\ \hline 12 = 4x \end{array}$$

Step 2: Undo multiplication or division.

$$\frac{12}{4} = \frac{4x}{4}$$
$$3 = x$$

Let's try another one.

Solve: $4y - 8 = 2y + 12$

Step 1: Undo addition or subtraction twice, once for variables and once for numbers.

$$\begin{array}{r} 4y - 8 = 2y + 12 \\ -2y \quad\quad -2y \\ \hline 2y - 8 = 12 \end{array}$$

Helpful Hint!
When variables are on both sides, move the variables to the left of the equals sign. This will make solving the equation easier.

Undo addition or subtraction again!

$$\begin{array}{r} 2y - 8 = 12 \\ +8 \quad +8 \\ \hline 2y \quad = 20 \end{array}$$

Step 2: Undo multiplication or division.

$$\frac{2y}{2} = \frac{20}{2}$$
$$y = 10$$

PAINLESS TIP

Let's Review Our Steps!

1. Simplify (distribute and/or combine like terms).
2. Undo addition or subtraction (variables and numbers).
3. Undo multiplication or division.

Let's try one more!

Solve: $3(x + 2) + 4 = 2x + 3x + 20$

Step 1: Simplify (distribute, combine like terms).
$3(x + 2) + 4 = 2x + 3x + 20$
$3x + 6 + 4 = 2x + 3x + 20$

Step 2: Undo addition or subtraction.

$$3x + 10 = 5x + 20 \rightarrow \text{move the variables to}$$
$$\underline{-5x \qquad - 5x} \qquad \text{the left}$$
$$-2x + 10 = 20$$

$$-2x + 10 = 20 \rightarrow \text{move the numbers to the}$$
$$\underline{\quad -10\ -10} \qquad \text{right}$$
$$-2x = 10$$

Step 3: Undo multiplication or division.

$$\frac{-2x}{-2} = \frac{10}{-2}$$
$$x = -5$$

BRAIN TICKLERS   Set # 13

Solve.

1. $8x - 2 = 6x + 10$

2. $2x + 6x - 4 = 4x + 24$

3. $-2(x + 1) = 4x - 8$

4. $4(x - 3) = 5(x + 3)$

5. $5x + 3 = 14 - 6x$

(Answers are on page 55.)

More Multistep Equations

You have to be good at solving equations to be able to solve word problems. Let's practice a few more equations and then apply what you know to solve problems!

Look at these examples. Try to solve them yourself on a separate sheet of paper and use this as a check. Good luck!

Examples:

1. Solve: $6(2 + y) = 3(3 - y) + 2$

   Solution:
   $$6(2 + y) = 3(3 - y) + 2$$
   $$12 + 6y = 9 - 3y + 2$$
   $$12 + 6y = 11 - 3y$$
   $$\underline{ + 3y + 3y}$$
   $$12 + 9y = 11$$
   $$\underline{-12 - 12}$$
   $$\frac{9y}{9} = \frac{-1}{9}$$
   $$y = -\frac{1}{9} \quad \text{And yes, fractions are okay!}$$

2. Solve: $x + 5(x + 2) = x + 3(x + 6)$

   Solution: $x + 5(x + 2) = x + 3(x + 6)$

$$x + 5x + 10 = x + 3x + 18$$

$$6x + 10 = 4x + 18$$

$$\underline{-4x \qquad\quad -4x}$$

$$2x + 10 = 18$$

$$\underline{-10 \; -10}$$

$$\frac{2x}{2} = \frac{8}{2}$$

$$x = 4$$

3. Tracy and Kelly drive to a baseball game together. The tickets cost $10.50 each, and parking was $6. Popcorn costs $2.25 a box. Tracy and Kelly each bought a box of popcorn. Kelly bought 2 drinks, and Tracy bought 3 drinks. They spent a total of $39.75 at the game between tickets, parking, and food. How much did each drink cost?

Solution:
Before you attempt this problem, read it one sentence at a time! Write down what you know. Remember, you always know something!

Tickets = $10.50 and you need 2

Parking = $6

Popcorn = $2.25 and you need 2

Drinks = ?? call this x. And Kelly bought 2 ($2x$) and Tracy bought 3 ($3x$)

Total = $39.75

Turn this into an equation!

$$10.50(2) + 6 + 2.25(2) + 2x + 3x = 39.75$$
$$21 + 6 + 4.50 + 5x = 39.75$$
$$31.50 + 5x = 39.75$$
$$\underline{-31.50 \qquad\quad -31.50}$$
$$\frac{5x}{5} = \frac{8.25}{5}$$
$$x = 1.65$$

Each drink cost $1.65.

Check! $10.50(2) + 6 + 2.25(2) + 2(1.65) + 3(1.65)$
$21 + 6 + 4.50 + 3.30 + 4.95$
39.75 Check!

4. The square and the equilateral triangle have the same perimeter. If the side of the square is $x + 5$, and one side of the equilateral triangle is $3x$, find the perimeter of each shape.

Solution:

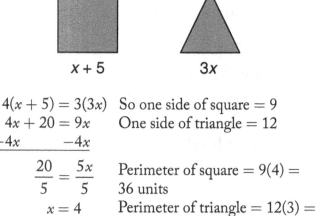

$x + 5$ $\qquad\qquad\qquad$ $3x$

$4(x + 5) = 3(3x)$ $\quad$ So one side of square $= 9$
$4x + 20 = 9x$ $\qquad$ One side of triangle $= 12$
$\underline{-4x \qquad\quad -4x}$

$\frac{20}{5} = \frac{5x}{5}$ $\qquad$ Perimeter of square $= 9(4) =$
$\qquad\qquad\quad$ 36 units
$x = 4$ $\qquad$ Perimeter of triangle $= 12(3) =$
$\qquad\qquad\quad$ 36 units

Do they have the same perimeter as stated in the question? Yes, they do!

BRAIN TICKLERS   Set # 14

Solve each of the following.

1. Chris is choosing between two cell phone plans that offer the same amount of free minutes. Plan A charges $39.99 per month with additional minutes costing $0.45 per minute. Plan B costs $44.99 with additional minutes costing $0.40 per minute. How many additional minutes will it take for the two plans to cost the same amount?

2. The sum of three numbers is 161. The second number is 6 times the first, and the third number is 5 more than the second. Find the numbers.

3. Two sides of a triangle are equal in length. The length of the third side is three more than the length of the equal sides. The perimeter of the triangle is 93 centimeters. Find the sides of the triangle.

4. Is -3 a solution to the equation $3x - 5 = 4 + 2x$? Explain your reasoning.

5. In a triangle, the measure of the first angle is six times a number. The measure of the second angle is nine less than the first angle. The measure of the third angle is three times the number plus the measure of the first angle. Find the measure of each angle of the triangle.

(Answers are on page 56.)

0, 1, and Infinite Solutions

Until now, equations you have solved all have one answer. But that is not the only answer you can get when you solve equations. We will look at different types of solutions in this section and come back to this topic after you review how to graph lines!

The following examples show the three different solutions you can get when you solve a linear equation.

Case 1: One solution

This is the one you are most familiar with. Any equation you have solved up until this point ends with $x =$ some number. That is the only number that will work in the equation. Here is one more example to practice our equation solving skills!

$$-3(x + 5) + 2(2x + 1) = 27$$
$$-3x - 15 + 4x + 2 = 27$$
$$x - 13 = 27$$
$$x = 40 \longrightarrow \text{One solution: 40 is the}$$

One solution: 40 is the only number that will work in this equation.

Case 2: Infinite solutions

What does it mean if $5 = 5$? It is true all the time. 5 is always equal to 5.

Notice this equation: $x + 2 = x + 2$

Think about what this means. Any number added to two is equal to the same number added to 2. Easy enough, right?

Try it! $3 + 2 = 3 + 2, 5 + 2 = 5 + 2$

Now look at this equation: $2x + 5 = x + x + 5$

They look a little different, but remember what tools you have to solve equations! Combine like terms!

$2x + 5 = x + x + 5$ is the same as $2x + 5 = 2x + 5$.

What do you notice? They are the same equation. Therefore, again, ANY number you put in for x will be true because both sides of the equation are equal.

Solve this equation:

$$2(x + 4) + 5 = 4x - 2x - 7 + 20$$
$$2x + 8 + 5 = 2x + 13$$
$$2x + 13 = 2x + 13 \longrightarrow \text{They are the same equation!}$$
$$\underline{-13 \qquad\quad -13}$$
$$\frac{2x}{2} = \frac{2x}{2} \longrightarrow \text{They are still the same equation if you simplify.}$$
$$x = x \qquad\qquad \text{Still equal!}$$

The solution: All real numbers or infinite solutions

Any number will work in this equation because both sides are the same!

You may state your solution of **infinite solutions,** as soon as you realize (prove) that two sides of the equation are equal.

Case 3: No solution

Does $2 = 5$? NO! You know that and understand that. Does $5 + 1 = 7 + 3$? No! Both sides of the equation are not equal. This is false. It doesn't work! Let's look at an algebraic equation. Now look at this equation:

$$x + 3 = x + 5$$

Can a number plus 3 equal (be the same as) the same number plus 5? NO! Pick a number and try it; $2 + 3 \neq 2 + 5$. When this situation happens, there is **no solution.** No number will work.

Solve:

$$2(x + 1) = 2(x + 3)$$
$$2x + 2 = 2x + 6 \qquad\qquad \text{Notice by "inspection" these are not the same.}$$
$$\underline{-2x \qquad\quad -2x}$$
$$2 = 6 \qquad\qquad\quad \text{If you simplify, it is still not the same. Not true!}$$

Therefore, the solution to this equation is **no solution**.

Anytime one side does not equal the other, there is no solution.

Let's summarize!

Types of solutions	Characteristics	Examples
One solution	Equation can be solved for a single variable.	$2x = 8$ or $2x + 6 = 12$
No solution	Variables cancel out.	$6 \neq 4$
	Constants are not equal.	$0 \neq 12$
	Once simplified, there is not a value that will make the equation true.	
Infinite solutions	Once simplified, there are equal values on both sides of the equal sign.	$5 = 5$ $x = x$ $3m = 3m$

 BRAIN TICKLERS   Set # 15

Determine if the equations have one, none, or infinite solutions. Justify your answer.

1. $5(8x + 4) + 7 = 267$

2. $4(8x - 1) = 19 + 32x$

3. $-5(x - 1) = 5 - 5x$

4. $12 + 4n = 4(n + 3)$

5. $-11 + x = -7x + 8x + 8$

6. $-21 - 8x = -1 + 6(4 - 5x)$

(Answers are on page 56.)

BRAIN TICKLERS—THE ANSWERS

Set # 9, page 36

1. $x = 3$

2. $x = 2$

3. $x = -4$

4. $x = 28$

5. $x = 24$

6. \$54

7. -16

8. 15 rides

Set # 10, page 38

1. $17x$

2. $4y - 8$

3. $a + 18$

4. $8c + 8d + 8$

5. $6d + 5e + 23$

Set # 11, page 40

1. $5y + 10$

2. $6y + 24$

3. $4x - 28$

4. $15x - 3$

5. $6x$

6. $5x + 15$

Set # 12, page 45

1. $x = 12$

2. $x = 2$

3. $x = 15$

4. $x = 3$

5. $x = 16$

6. $x = -2$

7. $x = 10$

Set # 13, page 48

1. $x = 6$

2. $x = 7$

3. $x = 1$

4. $x = -27$

5. $x = 1$

Set # 14, page 51

1. 100 minutes
2. 12, 72, 77
3. 30, 30, 33
4. No, $(x = 9)$ or $-14 \neq -2$
5. $45°, 54°, 81°$

Set # 15, page 54

1. One solution $(x = 6)$
2. No solution
3. Infinite solutions
4. Infinite solutions
5. No solution
6. One solution $(x = 2)$

Chapter 3

Linear Equations

Graphing a Line Using a Table of Values

If you can graph points, you can graph a straight line. Setting up an x-y chart (also known as a function table) will make graphing painless. Follow these easy steps.

Step 1: Solve the equation for y.

Step 2: Choose at least three values (input values) for x to help you make your chart.

Step 3: Substitute each value for x to find the value of y (output value).

Step 4: Graph the three points (x, y).

Step 5: Connect these three points with a straight line, and put arrows on the line.

1+2=3 MATH TALK!

When choosing x-values, choose a negative, zero, and a positive. This will cover every section of the graph paper.

Make sure you extend the line past the three points and put arrows on the ends. This indicates the line goes on forever.

Example 1:

Graph the line $y = x + 3$.

Step 1: The line is already in the form $y =$.

Step 2: Choose three values for x.

Step 3: Substitute each value for x and solve for y.

x	$x + 3$	y
-2	$-2 + 3$	1
0	$0 + 3$	3
2	$2 + 3$	5

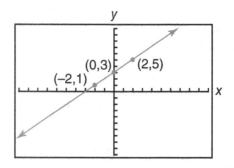

Step 4: Plot the points $(-2, 1)$, $(0, 3)$, and $(2, 5)$.

Step 5: Connect the three points with a straight line, and put arrows on the line.

1+2=3 MATH TALK!

When using a table of values to graph lines, it is not necessary to write the equation in $y =$ form (function form). However, this form does make graphing the line easier.

Example 2:

Graph the line $y - x = 5$.

Step 1: Solve the equation for $y =$. $\quad y - x = 5$
Add x to both sides.
$$\underline{+x \quad +x}$$
$$y = 5 + x$$
$$(\text{or } y = x + 5)$$

Step 2: Choose three values for x.

Step 3: Substitute each value for x and solve for y.

x	5 + x	y
−3	5 + −3	2
0	5 + 0	5
3	5 + 3	8

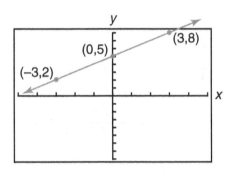

Step 4: Plot the points $(-3, 2)$, $(0, 5)$, and $(3, 8)$.

Step 5: Connect the three points with a straight line, and put arrows on the line.

Example 3:

Graph $6x + 2y = 12$.

Step 1: Solve the equation for $y =$. $6x + 2y = 12$
     → Subtract $6x$ $\underline{-6x \qquad\qquad -6x}$
        from both sides. $\dfrac{2y}{2} = \dfrac{-6x}{2} + \dfrac{12}{2}$

     → Since y has a coefficient
        of 2, divide by 2 to get
        y alone. Both -6 and 12
        must also be divided by 2. $y = -3x + 6$

Step 2: Choose three values for x.

Step 3: Substitute each value for x and solve for y.

x	$-3x + 6$	y
-2	$-3(-2) + 6$	12
0	$-3(0) + 6$	6
2	$-3(2) + 6$	0

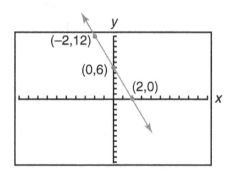

Step 4: Plot the points $(-2, 12)$, $(0, 6)$, and $(2, 0)$.

Step 5: Connect the three points with a straight line, and put arrows on the line.

CAUTION–Major Mistake Territory!

When solving for y, if y has a number in front of it (coefficient), when you divide y by the number to get y alone, you must also divide <u>EACH</u> term on the other side of the equation by that number!

Example:

Given: $3y = 12x + 15$

To solve for y, you must divide by 3; therefore, divide EACH term by 3.

$$\frac{3y}{3} = \frac{12x}{3} + \frac{15}{3}$$

$$y = 4x + 5$$

BRAIN TICKLERS   Set # 16

If needed, solve each equation for y. Make a table of values, and graph the line.

1. $y = 2x + 1$

2. $x + y = -2$

3. $y - 5 = x$

4. $2x - y = 3$

(Answers are on pages 115–116.)

Graphing Horizontal and Vertical Lines

Picture the horizon or the floor that you walk on. These are examples of horizontal lines. Horizontal lines go from left to right and are **parallel to the x-axis**. Horizontal lines are written in the form $y = b$, where b is where the line crosses the y-axis. Examples of equations of horizontal lines are $y = 2$, $y = -3$, and $y = 0$. All points on the horizontal line have the same y-coordinate.

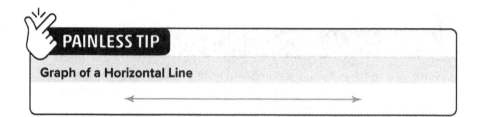

The walls in your house or a telephone pole represent vertical lines. Vertical lines go straight up and down and are parallel to the y-axis. A vertical line is written in the form $x = a$, where a is where the line crosses the x-axis. Examples of equations of vertical lines are $x = 2$, $x = -4$, and $x = 0$.

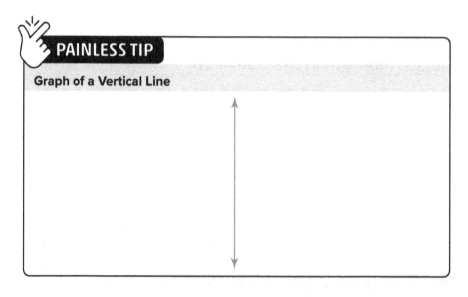

To graph horizontal and vertical lines, make a table of values.

Examples:

Graph $y = 2$.

For a table, this means the y-value always has to be 2. Choose three different x-values to make your table. No matter what x-values you choose, the y-value will always be 2. Then plot the points, and draw the line through the points. This is a horizontal line.

x	y
−2	2
0	2
3	2

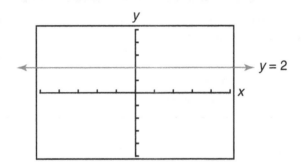

Graph $x = -1$.

Make a table, but this time the x-values always stay the same. The x-value has to be −1. Choose three different y-values to help you make the graph. Plot the points, and draw the line. This is a vertical line.

x	y
−1	−2
−1	0
−1	2

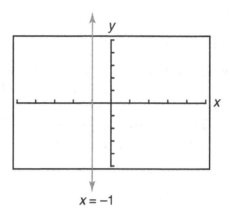

$$x = -1$$

CAUTION—Major Mistake Territory!

A horizontal line is always written in the form $y = b$. A vertical line is always written in the form $x = a$.

Examples:

$y = 3 \rightarrow$ Horizontal line

$x = -2 \rightarrow$ Vertical line

BRAIN TICKLERS   Set # 17

Graph each line.

1. $x = 4$

2. $x = -2$

3. $y = 5$

4. $y = -1$

(Answers are on pages 117–118.)

Slope of a Line

Have you ever tried to run up a steep hill? Or ride a bike all the way up a hill without walking? Or ski down a hill? The concept of "steepness" is one that can easily be understood in the real world.

Examples of different "slopes" are easy to see. As a beginner skier, which hill would you prefer to ski down?

Bunny Hill

Black Diamond

A beginner skier starts on the bunny hill, which is less steep than a hill considered a black diamond!

In the world of mathematics, the word **slope** describes the "steepness" of a line.

There are four basic types of slope.

Positive slope	Negative slope
Slants upward, from left to right	Slants downward, from left to right
Zero slope	No slope/Undefined slope
A horizontal line, left to right Has zero slope (like a floor)	A vertical line, up and down Has undefined slope (like a wall)

The slope of a line is described as a ratio. It is the comparison of the vertical change in the line compared to the horizontal change in the line. Here is the formal definition of slope.

$$\text{Slope} = \frac{\text{change in } y}{\text{change in } x} = \frac{\Delta y}{\Delta x}$$

For a *painless* way to remember slope, use this saying:

$$\text{Slope} = \frac{\text{Rise}}{\text{Run}}$$

- **Rise** is the vertical distance between points. Rise involves the *y*-axis. A positive rise means move up; a negative rise means move down.
- **Run** is the horizontal distance between two points. When you move to the right, it is a positive run; moving to the left is a negative run.

Realize that slope can also be the **constant rate of change**. Why? Think of a staircase or a ramp; the steepness changes by the same amount (constant) all the way up or down.

You can find slope of a line using two different methods. It will depend on if you are looking at a graph, an equation, or a table.

Method 1
Graph the two points and find the slope using rise over run.

Method 2
Use the slope formula when given two points.

$$m = \frac{y_2 - y_1}{x_2 - x_1}$$

Method 1—Finding slope from a graph
Find the slope of the following lines.

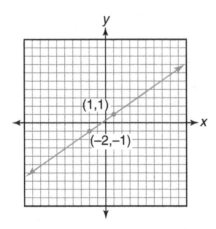

What does slope mean? $\dfrac{\text{Rise}}{\text{Run}}$

Start at the lowest point, $(-2, -1)$. Rise up 2 (positive); run right 3 (positive).

$$\text{Slope} = \frac{2}{3}.$$

1+2=3 MATH TALK!

If you always start with the lowest point, you will rise in a positive direction every time.

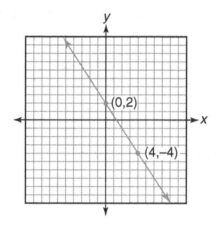

$$\text{Slope} = \frac{\text{Rise}}{\text{Run}}$$

Start at the lowest point $(4, -4)$. Rise up 6 (positive); run left 4 (negative).

$$\text{Slope} = \frac{6}{-4}.$$

The slope can reduce to $\dfrac{3}{-2}$, or $-\dfrac{3}{2}$.

CAUTION—Major Mistake Territory!

Reducing the slope does not change your answer. You will still plot a point on the same line!

Slope $= \dfrac{4}{6}$. Slope $= \dfrac{2}{3}$.

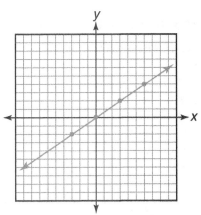

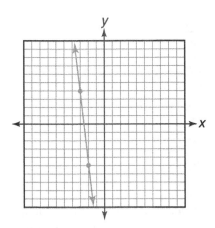

Start at the lowest point $(-2, -5)$.
Rise up 9 (positive) to get to the other point.
Run left 1 (negative) to get to the other point.

$$\text{Slope} = \frac{9}{-1}.$$

$$\text{Slope} = -9.$$

> **1+2=3** **MATH TALK!**
>
> $\dfrac{9}{-1}$ is the same as $\dfrac{-9}{1}$
>
> Moving up 9 and left 1 from the lowest point will get you to the same place as moving down 9 and right 1 when starting from the highest point.

Method 2—Using the formula

To find the slope of a line when given two points, use the formula

$$m = \frac{\text{rise}}{\text{run}} = \frac{y_2 - y_1}{x_2 - x_1}$$

What this formula means is that you find the difference in the y-coordinates (subtract the y-values) and divide by the difference of the x-coordinates (subtract the x-values).

It is important to remember that for the coordinate $(5, 6)$, 5 is the x-coordinate and 6 is the y-coordinate.

Find the slope of a line passing through the points $(4, 3)$ and $(5, 6)$.

$$m = \frac{y_2 - y_1}{x_2 - x_1}$$

$$m = \frac{6 - 3}{5 - 4}$$

$$m = \frac{3}{1}$$

The slope equals 3.

Always have the y-values in the numerator. Also, you have to subtract the x-values in the same order that you subtracted the y-values. What does this mean? Whichever y-value comes first on top, the x-value from the <u>SAME COORDINATE PAIR</u> has to come first as well, for the bottom of the fraction.

Example 1:

Find the slope of a line passing through the points $(-2, 4)$ and $(5, 6)$.

$$m = \frac{y_2 - y_1}{x_2 - x_1}$$

$$m = \frac{4 - 6}{-2 - 5} \text{ or } m = \frac{6 - 4}{5 - (-2)}$$

$$m = \frac{-2}{-7} \text{ or } m = \frac{2}{7}$$

The slope equals $\frac{2}{7}$.

Let's try one more to make sure we have it!

Example 2:

Find the slope of a line passing through the points $(4, -3)$ and $(-5, 2)$.

$$m = \frac{y_2 - y_1}{x_2 - x_1}$$

$$m = \frac{-3 - 2}{4 - (-5)} \text{ or } m = \frac{2 - (-3)}{-5 - 4}$$

$$m = \frac{-5}{9} \text{ or } m = \frac{5}{-9}$$

$$m = -\frac{5}{9}$$

CAUTION—Major Mistake Territory!

When using the slope formula, make sure you always have the *y*-values on top!

(5, 7) and (4, 3)

$$m = \frac{y_2 - y_1}{x_2 - x_1} = \frac{3 - 7}{4 - 5} = \frac{-4}{-1} = 4 \text{ or } \frac{7 - 3}{5 - 4} = \frac{4}{1} = 4$$

Finding slope from a table

You can also use the slope formula to find slope from a table.

Example 1:

The table shows the number of pages Sue has left to read after a certain number of minutes. The points lie on a line. Find the slope of the line, and interpret its meaning in context of the problem.

Time in minutes (x)	Pages left (y)
1	12
3	9
5	6

To find the slope, choose any two points from the table. Since a line has a constant rate of change (slope), the slope will be the same for any two points you choose. For this one example, slope will be shown using all of the points.

$$m = \frac{y_2 - y_1}{x_2 - x_1}$$

(1, 12) and (3, 9)   (3, 9) and (5, 6)   (1, 12) and (5, 6)

$$\frac{9 - 12}{3 - 1}$$ $$\frac{6 - 9}{5 - 3}$$ $$\frac{6 - 12}{5 - 1}$$

$$\frac{-3}{2}$$ $$\frac{-3}{2}$$ $$\frac{-6}{4} = \frac{-3}{2}$$

The slope of the line is $\frac{-3}{2}$. What does this mean in terms of the problem? For every 3 pages that are read, 2 minutes has gone by.

You can also see the slope in the table.

Time in minutes (x)	Pages left (y)
1	12
3	9
5	6

Remember, slope is the change in y over the change in x. Look down the page column (y): It decreases by 3 every time, hence, a change of -3. And the time column (x) increases by 2 every time, showing a change of 2. Again, this shows that the change in y over the change in x is $\dfrac{-3}{2}$.

Example 2:

Find the slope.

x	−6	−2	2
y	−2	−1	0

Solution:
Choose two points: $(-6, -2)$ and $(-2, -1)$. Use the slope formula.

$$m = \frac{y_2 - y_1}{x_2 - x_1}$$

$$\frac{-1 - (-2)}{-2 - (-6)}$$

$$= \frac{-1 + 2}{-2 + 6}$$

$$= \frac{1}{4}$$

The slope is $\dfrac{1}{4}$. What does this mean? For every 1 unit y increases, x increases by 4.

BRAIN TICKLERS   Set # 18

Find the slope of each line.

1.

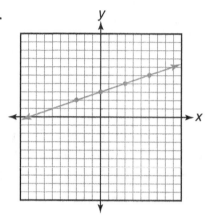

2.

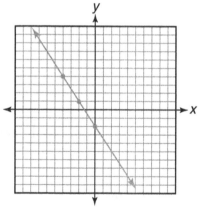

3.

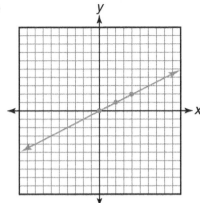

4.

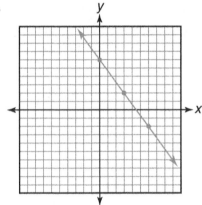

Find the slope of the line passing through the given points.

5. (5, 3) and (8, 2)

6. (−4, −4) and (5, 8)

7. (−3, 8) and (−2, 6)

8.

x	y
2	−3
4	0
6	3
8	6

9.

x	y
1	6
2	6
3	6
4	6

10. The table shows the number of gallons of paint Frank used to paint rooms in his house. Find the slope and interpret its meaning.

Gallons	2	4	6	8
Rooms painted	3	6	9	12

(Answers are on page 119.)

Slope of Horizontal and Vertical Lines

Slope—Two special cases

Let's review the four main types of slope.

Positive slope	Negative slope
Slants upward, from left to right	Slants downward, from left to right
Zero slope	No slope/Undefined slope
A horizontal line, left to right Has zero slope (like a floor)	A vertical line, up and down Has undefined slope (like a wall)

So far, you have learned how to find the slope of a positive or negative line. But what if the lines are horizontal or vertical?

Horizontal Lines

A horizontal line has no "steepness" to it. It does not slant up or down; therefore, it does not have a positive or negative slope. A horizontal line has a slope of zero.

A nice trick to remember this when looking at the horizontal line is to think "what letter can I make out of this"? A Z for ZERO!! Look!

Horizontal line   Makes a Z → Slope = zero!

Let's prove that a horizontal line has a slope of zero by using our slope formula.

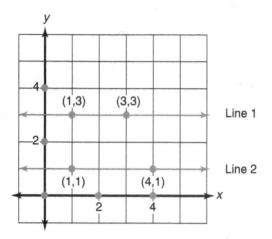

Let's use the coordinates of Line 1: $(1, 3)$ and $(3, 3)$.

$$m = \frac{y_2 - y_1}{x_2 - x_1}$$

$$m = \frac{3 - 3}{3 - 1}$$

$$m = \frac{0}{2}, \text{ which equals } 0!$$

Now use the coordinates of Line 2: $(1, 1)$ and $(4, 1)$.

$$m = \frac{y_2 - y_1}{x_2 - x_1}$$

$$m = \frac{1 - 1}{4 - 1}$$

$$m = \frac{0}{3}, \text{ which equals } 0!$$

The slope of a horizontal line is zero. Why is this true? Since a horizontal line never moves up or down, its y-coordinate will never change. This means the "change in y-coordinates" is always 0. When we divide by the change in x-coordinates to find the slope, our final answer will always be 0.

Vertical Lines

A vertical line is straight up and down, like a wall. The slope of a vertical line is undefined. A nice trick to help you remember the slope of a vertical line is to think of a letter this line can make. A U for undefined!

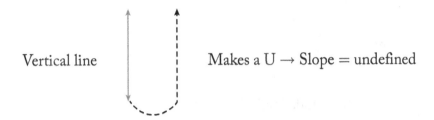

Vertical line   Makes a U → Slope = undefined

What does **undefined** mean? In math, it just means that the denominator of the fraction has a zero in it. Let's prove the slope of a vertical line is undefined by finding the slope between two points on a vertical line.

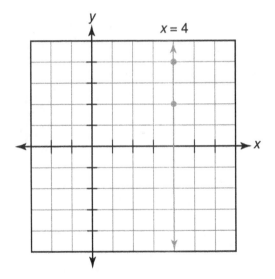

Let's use two points on the line to find the slope of the vertical line $x = 4$.

Use $(4, 2)$ and $(4, 4)$.

$$m = \frac{y_2 - y_1}{x_2 - x_1}$$

$$m = \frac{4 - 2}{4 - 4}$$

$$m = \frac{2}{0} \rightarrow \text{undefined}$$

We can't divide by zero. Any number divided by zero is undefined; therefore, **the slope of a vertical line is undefined**. Whenever you try to find the slope of a vertical line, you will get a number divided by zero. Therefore, the slope of every vertical line is undefined!

BRAIN TICKLERS   Set # 19

Find the slope of each line. State whether the line has a slope of zero or is undefined. Also state whether it is a horizontal or vertical line. Graph and label each line.

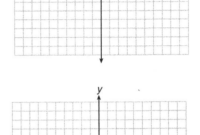

1. A (6, 3) and B (−5, 3)

2. C (1, −4) and D (7, −4)

3. E (3, 6) and F (3, −5)

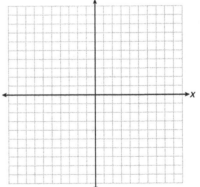

4. G (0, 4) and H (10, 4)

5. J (−4, 1) and K (−4, 7)

6. L (4, 0) and M (4, 10)

7. Explain how you can tell if a line is going to be horizontal or vertical just by looking at the coordinates. Give an example for each, and explain your reasoning.

(Answers are on pages 119–120.)

Applications of Slope

Did you know that when you buy a movie ticket for $10 a person, you could call that a constant rate of change or slope? Why, you ask? Look at the example below.

Example 1:

You and some friends go to the movies. The tickets cost $10 per person. The table below represents the number of people buying tickets and the total cost of the tickets.

Number of tickets (x)	Total cost ($) (y)
1	10
2	20
3	30
4	40
9	90
10	100

+1 ⟶ (between 1 and 2)   +10
+1 ⟶ (between 2 and 3)   +10

Remember that slope is the **change in y over the change in x**. Looking at each column, it is easy to see the change in y and change in x. Notice the change in y is 10 and the change in x is 1. Hence, the slope is 10/1.

Notice the table skips from 4 to 9, but if you were to continue the counting from 4, 5, 6, 7, 8, and 9, the pattern would be the same.

What does the slope mean in context of the problem? It costs $10 a ticket per person. Who knew! Slope shows up even at the movies!

Example 2:

Cole downloaded a bunch of songs from the Internet. His data is shown in the table below.

Time in minutes (x)	0	4	8	12
Number of songs (y)	0	8	16	24

a. What is the constant rate of change or slope? Explain what this means in context of the problem.
b. If Cole downloaded 36 songs, how many minutes would it take?
c. Graph the ordered pairs from the table on the graph below. From the graph, find the slope between two points using rise over run. What do you notice?

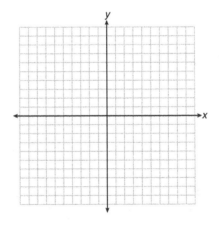

Solutions:

a. The slope is $\frac{2}{1}$. You can find this by noticing the change in y over the change in x or picking two points and using the slope formula. For points $(4, 8)$ and $(12, 24)$: $\frac{24 - 8}{12 - 4} = \frac{16}{8} = \frac{2}{1}$ → It takes 2 minutes to download 1 song.

b. Continue the pattern. The number of songs is always twice the number of minutes. So it would take 18 minutes to download 36 songs.

c.

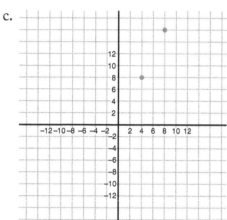

The graph is a line with a positive slope. It also has a slope of $\frac{2}{1}$.

SLOPE TOOLBOX!

When solving problems using slope, these are the main concepts of slope you need to remember:

$$\text{Slope} = \frac{\text{Rise}}{\text{Run}}$$ Use when graphing

$$\text{Slope} = \frac{\text{Change in } y}{\text{Change in } x} = \frac{\Delta y}{\Delta x}$$ Use when you have a table

$$m = \frac{y_2 - y_1}{x_2 - x_1}$$ Use when given two points or a table

BRAIN TICKLERS   Set # 20

1. The cost of having the vet make a house call for a pet is listed below.

Number of hours at house	0	1	2	3	5
Cost ($)	65	95	125	155	215

a. What is the vet charging per hour?

b. What is the fee for just calling the vet?

2. A balloon is released from the top of a building. The graph below shows the relationship of height over time of the balloon.

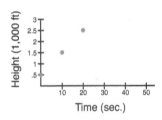

a. What is the starting height of the balloon?

b. How high was the balloon traveling per second?

c. If the balloon continued to rise at this constant rate, how high would the balloon be in 1 minute?

3. The pitch of a roof rises 6 inches for every 12 inches of horizontal distance it covers. What is the rate of change?

4. The table below shows John's savings account balance over the course of 6 months.

Month	0	1	2	3	4	5	6
Total	500	575	650	725	800	875	950

a. Find the rate of change using two different methods.

b. How much money did John first deposit into the bank?

c. If he continues to save at the same rate, how much money will he have saved in one year?

(Answers are on page 120.)

Slope-Intercept Form of a Line

Now that you are a pro at finding the slope of a line using points on a graph, let's see how to find the slope and y-intercept of a line by looking at its equation.

The equation of a line can be written in the form $y = mx + b$.

$$y = mx + b$$

The number in front of x is the slope (m)

b (including the sign) is the y-intercept

When given the equation, you can find the slope and y-intercept.

$y = 2x + 5$

Slope $= 2$ y-intercept $= 5$

$y = -3x - 2$

Slope $= -3$ y-intercept $= -2$

1+2=3 MATH TALK!

Remember that Slope $= \dfrac{\text{Rise}}{\text{Run}}$.

A slope of 2 is the same as a slope of $\dfrac{2}{1}$.

Writing slope as a fraction is helpful when graphing.

If an equation is not written in slope-intercept form, solve for y and rewrite it in that form.

Example:

Find the slope and y-intercept of the equation $2x + 4y = 8$.
This equation is not in slope-intercept form. Let's change that!

$$2x + 4y = 8$$

Subtract $2x$ from both sides. $\underline{-2x \qquad\qquad -2x}$

Divide both sides by 4. $\dfrac{4y}{4} = \dfrac{-2x}{4} + \dfrac{8}{4}$

Simplify. $y = -\dfrac{1}{2}x + 2$

The equation written in slope-intercept form is

$$y = -\dfrac{1}{2}x + 2.$$

It is easy to see that the slope is $-\dfrac{1}{2}$ and the y-intercept is 2.

Exciting examples:

Now see if you can find the slope and y-intercept. The
answers are listed on the next page. Cover them up and test
yourself. No cheating!

1. $y = \dfrac{3}{4}x - 8$

   Slope = _____

   y-intercept = _____

2. $y = -x + 1$

   Slope = _____

   y-intercept = _____

3. $y = 4x$

   Slope = _____

   y-intercept = _____

4. $y = 2$

   Slope = _____

   y-intercept = _____

5. $2y = 6x - 10$

   Slope = _____

   y-intercept = _____

Answers:

1. Slope $= \dfrac{3}{4}$
   y-intercept $= -8$

2. Slope $= -1$
   y-intercept $= 1$

3. Slope $= 4$
   y-intercept $= 0$

4. Slope $= 0$
   y-intercept $= 2$

5. Slope $= 3$
   y-intercept $= -5$

For question 5, make sure you place the line into slope-intercept form, $y = mx + b$. The original equation $2y = 6x - 10$ then becomes $y = 3x - 5$.

More examples:

6. Fun Boat Marina charges a one-time $25 rental fee for a boat in addition to charging $15 per hour for usage. The total cost of renting the boat, y, for x hours can be represented by the equation $y = 15x + 25$. State the slope and y-intercept and explain what they mean in the context of the problem.

   Solution:
   Slope $= 15$, which means they charge $15 per hour to rent the boat; y-intercept $= 25$, which represents a one-time rental fee.

7. Jim is reading a 250-page book. He can read 40 pages in an hour. The equation for the number of pages he has left to read is $y = 250 - 40x$, where x is the number of hours he reads.

   a. State the slope and y-intercept of the equation.

   b. Interpret what the slope and the y-intercept represent.

   Solutions:
   a. Slope $= -40/1$ and the y-intercept $= 250$.

   b. Slope $= -40/1$ means he reads 40 pages per hour. It is negative because for every hour he reads, he has 40 fewer pages to read in the book. The y-intercept of 250 represents the starting number of pages in the book.

BRAIN TICKLERS   Set # 21

Find the slope and y-intercept of each line. Make sure the line is in the form $y = mx + b$.

1. $y = -3x + 1$

2. $2y = 4x + 8$

3. $y = -4 - x$

4. $x + y = 12$

5. $y = -2x$

6. $y = 8$

7. $y + 5x = 0$

8. $-2x + 9 = 3y$

9. Gaming and Skate Fun Zone charges an entrance fee of $12 plus $3.25 per hour to skate. The equation $y = 12 + 3.25x$ represents the total charge, y, in terms of the number of hours skating, x. State and interpret the slope and the y-intercept.

(Answers are on page 121.)

Graphing Using the Slope-Intercept

Now that you are a pro at finding the slope and y-intercept of a line, let's graph lines using this information.

1+2=3 MATH TALK!

Equation of a line: $y = mx + b$, where $m = $ slope and $b = y$-intercept

$$\text{Slope} = \frac{\text{Rise}}{\text{Run}}$$

y-intercept $=$ where the line crosses the y-axis

To graph a line using the slope-intercept method, follow these easy steps.

Step 1: Write the equation in the form $y = mx + b$.

Step 2: State the slope.

Step 3: State the y-intercept.

Step 4: Graph the point of the y-intercept. (This will be on the y-axis.)

Step 5: From this point, do $\dfrac{\text{Rise}}{\text{Run}}$ to find and graph a second point.

Step 6: Draw a line to connect the two points.

Step 7: Label the line.

Example 1:

Graph the line $y = \dfrac{1}{2}x - 3$ using the slope-intercept method.

$\text{Slope} = \dfrac{1}{2}$

$y\text{-intercept} = -3$

Start at $(0, -3)$ because you are on the y-axis.

From $(0, -3)$, go up 1, right 2, and plot a point. From this point, go up 1, right 2 again, and plot a second point.

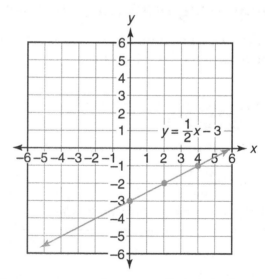

Draw a line connecting the points. Label the line with the equation $y = \dfrac{1}{2}x - 3$.

CAUTION–Major Mistake Territory!

The *y*-intercept is always on the *y*-axis. Therefore, the coordinate is always (0, *b*).

Example 2:

Graph the line $y = x + 1$ using the slope-intercept method.

Slope $= 1$

y-intercept $= 1$

Start at $(0, 1)$.

From $(0, 1)$, go up 1, right 1, and plot a point. Repeat up 1, right 1, and plot a second point.

Draw a line connecting the points. Label the line with the equation $y = x + 1$.

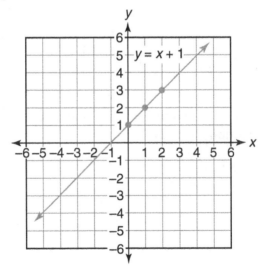

Example 3:

Graph $y = -2x$ using the slope-intercept method.

Slope $= \dfrac{-2}{1}$

y-intercept $= 0$; notice that this time there is no number after the x-term.

Start at $(0, 0)$ this time for the y-intercept.

From $(0, 0)$, go down 2, right 1, and plot a point. From this point, go down 2, right 1 again, and plot a second point.

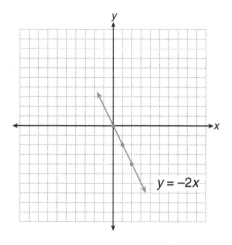

Draw a line connecting the points. Label the line with the equation $y = -2x$.

Example 4:

Graph the line $2x + 4y = 12$ using the slope-intercept method.

Remember, the line must be written in $y = mx + b$ form. Is this line in the correct form? No!

Rewrite in $y = mx + b$ form.

$$2x + 4y = 12$$
$$\underline{-2x \qquad\quad -2x}$$
$$\frac{4y}{4} = \frac{-2x}{4} + \frac{12}{4}$$

$$y = -\frac{1}{2}x + 3$$

$$\text{Slope} = -\frac{1}{2}$$

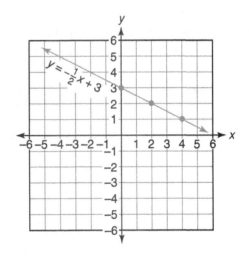

y-intercept $= 3$

Start at $(0, 3)$.

From $(0, 3)$, go down 1, right 2, and plot a point. From this point, go down 1, right 2, and plot a second point.

Draw a line connecting the points. Label the line with the equation $y = -\frac{1}{2}x + 3$.

BRAIN TICKLERS   Set # 22

Graph each line using the slope-intercept method.

1. $y = -3x - 1$

2. $y = \dfrac{1}{2}x + 2$

3. $x + y = -5$

4. $2x + 2y = 10$

5. A car rental charges a flat fee of $40 plus $12 per day. The function is represented by $c = 12d + 40$. State the slope of the y-intercept. What do they mean in the context of the problem? Graph the line.

(Answers are on pages 121–123.)

A Closer Look at $y = mx + b$

Now that you have reviewed how to graph a line using slope-intercept form, let's look at the equation a little bit more. Compare the equations in the two-column table below.

$y = mx$	$y = mx + b$
$y = 3x$	$y = 3x + 2$
slope $= 3$	slope $= 3$
y-intercept $= 0$	y-intercept $= 2$
Graph	Graph

$y = 3x$

$y = 3x + 2$

$y = mx$	$y = mx + b$
$y = -5x$	$y = -5x - 1$
slope $= -5$	slope $= -5$
y-intercept $= 0$	y-intercept $= -1$
Graph:	Graph:

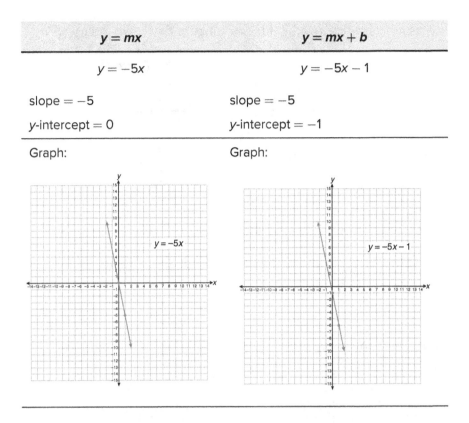

What do you notice?

Both equations are very similar. $y = mx + b$ has a slope and a y-intercept. $y = mx$ has a slope and a y-intercept of ZERO; hence, you never see the b.

It is important to remember that both of these equations graph lines (linear functions). But if a line graph passes through $(0, 0)$, it has a special name. It is called a **direct variation**. And this means the slope is also **proportional**.

Let's look at these two big ideas. Compare these two tables.

Table A	
x	**y**
1	3
2	6
3	9

Table B	
x	**y**
1	6
2	9
3	12

Let's find the slope of each table.

Table A: $(1, 3)$ and $(2, 6)$

$$\frac{6 - 3}{2 - 1} = \frac{3}{1}$$

Table B: $(1, 6)$ and $(2, 9)$

$$\frac{9 - 6}{2 - 1} = \frac{3}{1}$$

Both equations have the same slope. Now look at each graph.

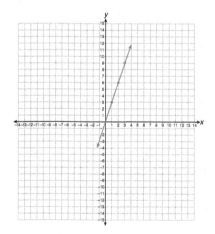

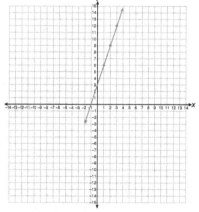

Table A passes
through $(0, 0)$.
Table A represents
a direct variation.

Table B does not pass
through $(0, 0)$.
Table B is still a line,
but NO direct variation.

You can also look at the tables to see if the points will reach $(0, 0)$.

Table A	
0	0
1	3
2	6
3	9

Table B	
0	3
1	6
2	9
3	12

If you follow the pattern for slope for each table, notice the tables also prove that Table A will pass through $(0, 0)$, but Table B will not.

Since Table A is a direct variation, it is also proportional. To determine if two quantities are proportional, compare the ratio of $\frac{y}{x}$ for several pairs of points.

Look at the tables again.

Table A	
x	y
1	3
2	6
3	9

Table B	
x	y
1	6
2	9
3	12

Look at EACH <u>pair</u> in Table A

$$\frac{y}{x} = \frac{3}{1} = \frac{6}{2} = \frac{9}{3}$$

Each ratio simplifies to $\frac{3}{1}$, showing this is proportional.

Look at EACH <u>pair</u> in Table B

$$\frac{y}{x} = \frac{6}{1} \neq \frac{9}{2} \neq \frac{12}{3}$$

Each ratio DOES NOT simplify to the same number. Therefore, it is NOT proportional.

1+2=3 MATH TALK!

If a line is a **direct variation**, it will pass through (0, 0), and it will be **proportional** (each pair will reduce to the same ratio).

KEY CONCEPT: You can use tables, words, equations, or graphs to compare proportional relationships.

TABLE:

x	15	20	25	30
y	3	4	5	6

$$\frac{y}{x} = \frac{3}{15} = \frac{4}{20} = \frac{5}{25} = \frac{6}{30}$$

WORDS: Constant of proportionality $= \frac{1}{5}$

EQUATION: $y = \frac{1}{5}x$ Slope $= \frac{1}{5}$, y-intercept $= 0$

GRAPH: will pass through (0, 0)

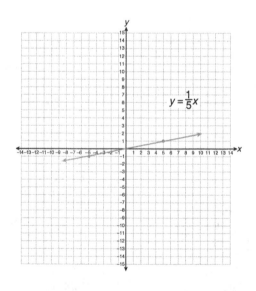

CAUTION—Major Mistake Territory!

When checking for proportionality, make sure you always do $\frac{y}{x}$ for each pair of points!

BRAIN TICKLERS   Set # 23

Determine whether each linear function is a direct variation. If so, state the constant of proportionality.

1.

Pictures (x)	5	6	7	8
Profit (y)	20	24	28	32

2.

Age (x)	10	11	12	13
Grade (y)	5	6	7	8

3.

x	6	9	12
y	1	2	3

4.

x	3	4	5
y	6	8	10

5.

x	4	5	6
y	6	8	10

6.

x	5	6	7
y	20	4	60

In each case, *y* varies directly as *x*. Find the missing numbers.

7.

x	1	2	?
y	5	?	25

8.

x	4	8	?
y	6	?	15

9. Which equation is a direct variation: $y = 3x + 1$ or $y = 3x$? Explain your reasoning.

(Answers are on page 123.)

A Closer Look at x- and y-Intercepts

The **x-intercept** of a graph is where the graph crosses the *x*-axis. This graph crosses the *x*-axis at (6, 0), so the *x*-intercept is 6.

The **y-intercept** of a graph is where the graph crosses the *y*-axis. This graph crosses the *y*-axis at (0, 3), so the *y*-intercept is 3.

You can use the *x*- and *y*-intercepts to help you graph a line.

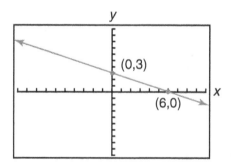

What if the problem says, "Graph the line with an *x*-intercept of −4 and a *y*-intercept of 2"? No problem!

The *x*-intercept is −4, so plot the coordinate (−4, 0) since it is on the *x*-axis. The *y*-intercept is 2, so plot the coordinate (0, 2) since it is on the *y*-axis. Then draw a line through the two points!

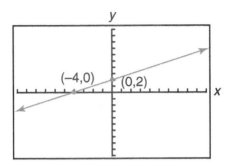

If you already have the equation of a line and want to find the x- and y-intercepts, you need to know two big things!

PAINLESS TIP

Finding the x- and y-Intercepts

To find the x-intercept, let $y = 0$ and solve for x.

To find the y-intercept, let $x = 0$ and solve for y.

Example 1:

Find the x- and y-intercepts of $2x + y = 10$.

x-intercept: Let $y = 0$, and solve for x.

$$2x + 0 = 10$$
$$\frac{2x}{2} = \frac{10}{2}$$
$$x = 5$$

The x-intercept is 5.
The coordinates of the x-intercept are $(5, 0)$.

y-intercept: Let $x = 0$, and solve for y.

$$2(0) + y = 10$$
$$0 + y = 10$$
$$y = 10$$

The y-intercept is 10.
The coordinates of the y-intercept are $(0, 10)$.

Now graph.

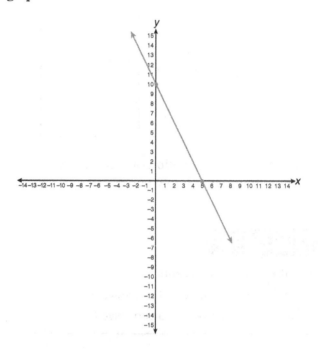

Example 2:

Find the x- and y-intercepts of $-3x + 6y = 18$.

x-intercept: Let $y = 0$, and solve for x.

$$-3x + 6(0) = 18$$
$$\frac{-3x}{-3} = \frac{18}{-3}$$
$$x = -6$$

The x-intercept is -6.
The coordinates of the x-intercept are $(-6, 0)$.

y-intercept: Let $x = 0$, and solve for y.

$$-3(0) + 6y = 18$$
$$\frac{6y}{6} = \frac{18}{6}$$
$$y = 3$$

The y-intercept is 3.
The coordinates of the y-intercept are $(0, 3)$.

You can use the *x*- and *y*-intercepts to graph the line. Plot the intercepts $(-6, 0)$ and $(0, 3)$. Then draw a line connecting the two points. Label the line with the equation.

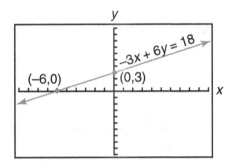

BRAIN TICKLERS   Set # 24

Find the *x*- and *y*-intercepts.

1. $2x + 4y = 20$

2. $2x - 10 = 2y$

3. $y = 2x + 8$

Find the *x*- and *y*-intercepts. Graph each line using the intercepts.

4. $x + 2y = 6$

5. $3x = 3y - 9$

(Answers are on page 124.)

Writing the Equation of a Line

There is more than one way to write the equation of a line. We will look at two different forms in this section:

- **Slope-intercept form**
- **Point-slope form**

The table below lists characteristics of each type.

Slope-intercept form	Point-slope form
$y = mx + b$	$y_2 - y_1 = m(x_2 - x_1)$
Given the slope and y-intercept, you can write the equation of the line.	**Given one point and a slope**, you can write the equation of the line.
Slope $= 2$, y-intercept $= -8$	Given (2, 3) and Slope $= 4$
Equation: $y = 2x - 8$	Equation: $(y - 3) = 4(x - 2)$

Here you will learn how to use each method.

Slope-intercept form

Think about what you know from graphing a line. If you were given $y = 2x + 4$, could you graph the line? Yes! You know the Slope $= 2$ and the y-intercept $= 4$.

Now, instead of the equation, you will get the information first. What is the equation of the line if the Slope $= 3$ and the y-intercept $= 5$?

Think about what you know about $y = mx + b$. Just plug it in! The equation of the line would be $y = 3x + 5$. Painless!

CAUTION—Major Mistake Territory!

For a problem to be an equation, it must include the equals sign!

Exciting examples:

Write the equation of the line in slope-intercept form.

1. Slope $= -2$, y-intercept $= 5$ $y = -2x + 5$
2. Slope $= \dfrac{1}{3}$, y-intercept $= -2$ $y = \dfrac{1}{3}x - 2$
3. Slope $= 4$, y-intercept $= 0$ $y = 4x$
4. Slope $= 0$, y-intercept $= 7$ $y = 7$

5. You have \$500 in the bank. You are going to start adding \$10 per week to your savings account. Write an equation to represent this situation.

   Slope $= 10$ (adding \$10 per 1 week)

   y-intercept $= 500$ (your initial amount)

   Equation: $y = 10x + 500$

6. It costs \$5 per ticket for a matinee movie. You also want to buy one box of popcorn for \$6. Write the equation of the line representing the total cost, y, based on the number of tickets purchased, x.

   Slope $= 5$ (\$5 per ticket)

   y-intercept $= 6$ (one-time fee for popcorn)

   Equation: $y = 5x + 6$

   a. If you and 10 friends went to the movies, how much would it cost?

   Solution:
   Substitute 11 for x.

   $y = 5x + 6$
   $y = 5(11) + 6$
   $y = 55 + 6$
   $y = 61$ It would cost \$61.

b. If someone spent \$131 at the movies, how many tickets did they buy?
This time substitute the total price \$131 for y.

$$131 = 5x + 6$$
$$\underline{-6 \qquad\qquad -6}$$
$$\frac{125}{5} = \frac{5x}{5}$$
$$x = 25 \qquad \text{The person bought 25 tickets.}$$

Point-slope form

If you are given a point and a slope, you can also write the equation of the line. Look at the following example:

Write the equation of the line that passes through the point $(1, 7)$ and has a slope of 3.

Solution:
Think about what you know. $(1, 7)$ represents the coordinates (x, y), and the slope of 3 represents the letter m for slope. Plug what you know into the point-slope formula.

POINT-SLOPE FORMULA: $y_2 - y_1 = m(x_2 - x_1)$

$$y - 7 = 3(x - 1)$$

You must keep one x and one y, so you still have an equation.

Examples:

Write the equation in point-slope form given the following:

1. Passes through $(4, 5)$ with a slope of -3.

$$y_2 - y_1 = m(x_2 - x_1)$$
$$y - 5 = -3(x - 4)$$

2. Passes through $(1, 3)$ with a slope of $\frac{1}{4}$.

$$y_2 - y_1 = m(x_2 - x_1)$$
$$y - 3 = \frac{1}{4}(x - 1)$$

3. Passes through $(-2, -5)$ with a slope of 4.

$$y_2 - y_1 = m(x_2 - x_1)$$
$$y - (-5) = 4(x - (-2))$$
$$y + 5 = 4(x + 2)$$

BRAIN BUSTER!

What if the problem said write the equation of the line that passes through the point (–2, 6) with a slope of 3 in slope-intercept form.

Solution:
You have a point (–2, 6) and a slope of 3.
Start with the *point-slope formula*.

$$y_2 - y_1 = m(x_2 - x_1)$$
$$y - 6 = 3(x - (-2))$$
$$y - 6 = 3(x + 2)$$

Now use what you know from simplifying the equation to get this into the final slope-intercept form.

$$y - 6 = 3(x + 2)$$
$$y - 6 = 3x + 6$$
$$\underline{+6 \qquad +6}$$
$$y = 3x + 12 \qquad \text{Slope-intercept form!}$$

Alternative method to solve the same problem

Let's write the equation of the line passing through the point $(-2, 6)$ with Slope $= 3$ starting with the *slope-intercept formula*.

You know $y = mx + b$. So plug in what you know, which is (x, y) and m. That will let you find b.

$$y = mx + b$$
$$6 = 3(-2) + b$$
$$6 = -6 + b$$
$$\underline{+6 \quad +6}$$
$$12 = b$$

If you know Slope $= 3$ and the y-intercept $= 12$, you can write the equation $y = 3x + 12$, which is the same equation as above!

More examples:

Write the equation of a line passing through the point $(-4, -7)$ with a slope of -1. Write the final answer in slope-intercept form.

POINT-SLOPE METHOD

$$y_2 - y_1 = m(x_2 - x_1)$$
$$y - (-7) = -1(x - (-4))$$
$$y + 7 = -1(x + 4)$$
$$y + 7 = -1x - 4$$
$$\underline{-7 \quad\quad -7}$$
$$y = -1x - 11$$

SLOPE-INTERCEPT METHOD

$$y = mx + b$$
$$-7 = -1(-4) + b$$
$$-7 = 4 + b$$
$$\underline{-4 \quad\quad -4}$$
$$-11 = b$$
$$y = -1x - 11$$

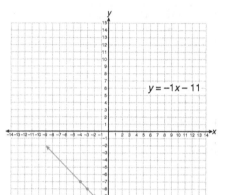

Write the equation of the line passing through the points $(2, 4)$ and $(6, 10)$.

This time you have a point, but you need a slope! You can find it because you know the slope formula.

$$\text{Find the slope} \quad m = \frac{y_2 - y_1}{x_2 - x_1}$$

$$\frac{10 - 4}{6 - 2} = \frac{6}{4}$$

$$\text{Slope} = \frac{3}{2}$$

Now use ONE point (it doesn't matter which one).

POINT-SLOPE METHOD

$(2, 4)$, Slope $= \dfrac{3}{2}$

$$y_2 - y_1 = m(x_2 - x_1)$$

$$y - 4 = \frac{3}{2}(x - 2)$$

$$y - 4 = \frac{3}{2}x - 3$$

$$\underline{+4 \qquad\qquad +4}$$

$$y = \frac{3}{2}x + 1$$

SLOPE-INTERCEPT METHOD

$(2, 4)$, Slope $= \dfrac{3}{2}$

$$y = mx + b$$

$$4 = \frac{3}{2}(2) + b$$

$$4 = 3 + b$$

$$\underline{-3 \qquad -3}$$

$$1 = b$$

$$y = \frac{3}{2}x + 1$$

BRAIN TICKLERS   Set # 25

Write the equation of each line, given the slope and y-intercept.

1. Slope $= 4$ y-intercept $= 10$

2. Slope $= \dfrac{1}{2}$ y-intercept $= 0$

3. Slope $= -5$ y-intercept $= -3$

4. Slope $= -\dfrac{2}{3}$ y-intercept $= 4$

5. Slope $= 0$ y-intercept $= 5$

Write the equation of the lines that pass through the following points, with the given slopes. Write the equation in point-slope form and y-intercept form.

6. Passes through the point $(-8, 2)$ and Slope $= 3$

7. Passes through the point $(4, 6)$ and Slope $= 2$

8. Passes through the point $(-6, -1)$ and Slope $= \dfrac{1}{3}$

9. Passes through the point $(5, -2)$ and Slope $= -4$

10. Write the equation of a line that passes through the points $(3, 6)$ and $(1, 10)$

11. Write the equation of a line that passes through the given two points

x	y
−2	−8
8	12

(Answers are on page 125.)

More writing equations of lines

Apply what you know about slope-intercept form and point-slope form to write an equation of the line for these problems.

Example 1:

A line has a slope of -3 and an x-intercept of 7. Write the equation for the line in slope-intercept form. Then write the equation of a line parallel to this line that has a y-intercept of -2.

Solution:
Remember that the coordinates of the x-intercept would be $(7, 0)$. You have a point and a slope given to you, so you can write the equation of the line.

$$y = mx + b$$
$$0 = -3(7) + b$$
$$0 = -21 + b$$
$$b = 21$$

The equation of the line would be $y = -3x + 21$.

For a line to be parallel to this line, the slopes have to be the same. So, the slope $= -3$, and the new y-intercept $= -2$. The new equation would be $y = -3x - 2$.

Example 2:

The math department sponsors a Family Fun Night each year. In the first year, there were 70 participants. In the third year, there were 114 participants.

a. Write an equation to represent this situation.

b. If this trend continues, how many participants can be expected in the sixth year?

Solution:

a. This problem is really giving you two coordinates. The 70 participants in the first year are represented by $(1, 70)$, and the 114 participants in the third year, $(3, 114)$.

Use these points $(1, 70)$ and $(3, 114)$ to write the equations.

$$\text{Slope} = \frac{114 - 70}{3 - 1} = \frac{44}{2} = 22$$

$$\begin{aligned}
y - y_1 &= m(x - x_1) \\
y - 70 &= 22(x - 1) \\
y - 70 &= 22x - 22 \\
\underline{+70 \qquad +70} & \\
y &= 48
\end{aligned}$$

Solution:

Equation is $y = 22x + 48$

b. Use the equation to make the prediction for year 6 (when $x = 6$).

$$\begin{aligned}
y &= 22x + 48 \\
y &= 22(6) + 48 \\
y &= 132 + 48 \\
y &= 180
\end{aligned}$$

If the pattern continues, they can expect 180 people for year 6.

Example 3:

All tickets for a country concert are the same price, and the ticket company charges a one-time user fee. A person who orders 5 tickets pays $385. A person who orders 3 tickets pays $235.

    a. Write an equation relating the total cost to the number of tickets purchased.

    b. If you spent $910, how many tickets did you buy?

Solution:

    a. Again, the problem is really giving you two points (5, 385) and (3, 235). Use these two points to write the equation of the line. This time the solution will show how to do this using the $y = mx + b$ method.

Given: (5, 385) and (3, 235)

$$\text{Slope} = \frac{235 - 385}{3 - 5} = \frac{-150}{-2} = 75$$

Now choose one point (5, 385) and use slope $= 75$ to find b.

$$y = mx + b$$
$$385 = 75(5) + b$$
$$385 = 375 + b$$
$$b = 10$$

Solution:

The equation of the line is $y = 75x + 10$.

    b. Set $y = 910$ and solve for x.

$$75x + 10 = 910$$
$$\underline{-10 \quad -10}$$
$$\frac{75x}{75} = \frac{900}{75}$$
$$x = 12 \qquad \text{You bought 12 tickets.}$$

BRAIN TICKLERS   Set # 26

1. A cab ride from the train station to the nearby hotel was 20 miles long. The total cost of the trip was $35.50. If the cab driver charges $1.25 per mile, write an equation that could be used to represent the cost of any cab ride.

2. A gym membership costs $208 for 6 months and $406 for 12 months. This includes a monthly fee plus a one-time activation fee.

   a. Write an equation that represents the total cost C, related to the months m.

   b. What would the cost be for an 18-month membership?

3. Write the equation of a line with Slope $= -4$ and y-intercept $= -2$.

4. If 8 pencils cost $3.35 and 10 pencils cost $4.25, write a linear equation to represent this function where y represents total cost (in dollars) and x represents the number of pencils.

(Answers are on page 125.)

BRAIN TICKLERS—THE ANSWERS

Set # 16, page 61

1. $y = 2x + 1$

x	2x + 1	y
−2	−4 + 1	−3
0	0 + 1	1
2	4 + 1	5

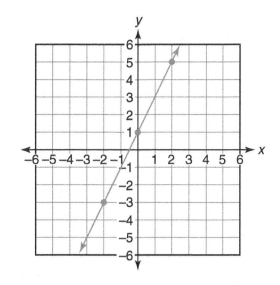

2. $x + y = -2$
$y = -x - 2$

x	−x − 2	y
−2	2 − 2	0
0	0 − 2	−2
2	−2 − 2	−4

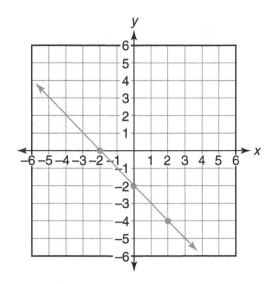

3. $y - 5 = x$
   $y = x + 5$

x	$x + 5$	y
-2	$-2 + 5$	3
0	$0 + 5$	5
2	$2 + 5$	7

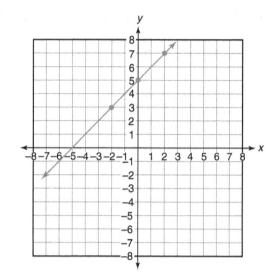

4. $2x - y = 3$
   $y = 2x - 3$

x	$2x - 3$	y
-2	$-4 - 3$	-7
0	$0 - 3$	-3
2	$4 - 3$	1

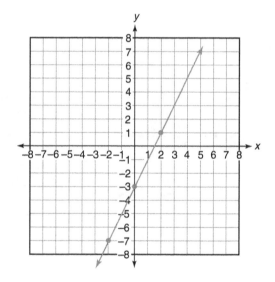

Set # 17, page 64

1. $x = 4$

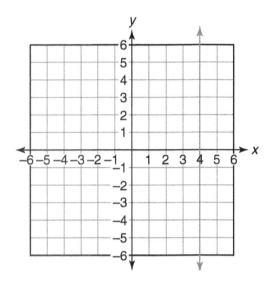

2. $x = -2$

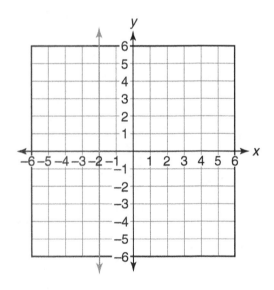

3. $y = 5$

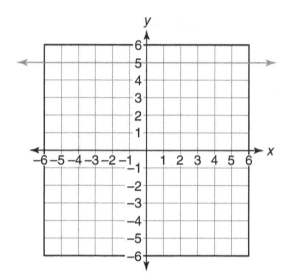

4. $y = -1$

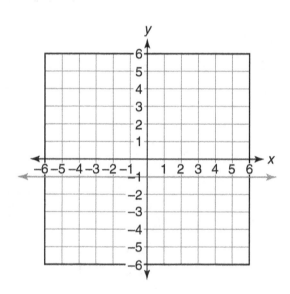

Set # 18, pages 74–76

1. $\dfrac{1}{3}$

2. $-\dfrac{3}{2}$

3. $\dfrac{1}{2}$

4. $-\dfrac{4}{3}$

5. $-\dfrac{1}{3}$

6. $\dfrac{4}{3}$

7. -2

8. $\dfrac{3}{2}$

9. 0

10. Slope $=\dfrac{3}{2}$. Frank can paint 3 rooms with 2 gallons of paint or Frank can paint 1.5 rooms with 1 gallon of paint.

Set # 19, page 81

1. Slope $= 0$, horizontal line

2. Slope $= 0$, horizontal line

3. Slope $=$ undefined, vertical line

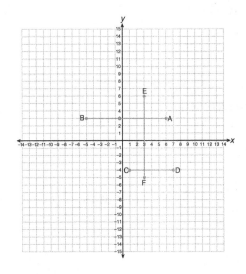

4. Slope $= 0$, horizontal line

5. Slope $=$ undefined, vertical line

6. Slope $=$ undefined, vertical line

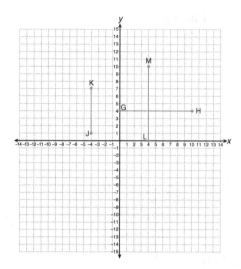

7. If the *x*-coordinates are the same, such as (2, 3) and (2, 5), it will be a vertical line. If the *y*-coordinates are the same, such as (1, 6) and (−3, 6), it will be a horizontal line.

Set # 20, page 85

1. a. $30 per hour
   b. $65

2. a. 500 feet
   b. 100 feet/second
   c. 6,500 feet

3. 6 inches/12 inches or $\dfrac{1}{2}$

4. a. $\dfrac{\text{change in } y}{\text{change in } x} = \dfrac{75}{1}$ or show work for slope formula
   b. $500
   c. $1,400

Set # 21, page 89

1. Slope $= -3$, y-intercept $= 1$

2. Slope $= 2$, y-intercept $= 4$

3. Slope $= -1$, y-intercept $= -4$

4. Slope $= -1$, y-intercept $= 12$

5. Slope $= -2$, y-intercept $= 0$

6. Slope $= 0$, y-intercept $= 8$

7. Slope $= -5$, y-intercept $= 0$

8. Slope $= -\dfrac{2}{3}$, y-intercept $= 3$

9. Slope $= 3.25$, they charge \$3.25 to rent per hour; y-intercept $=$ 12, represents \$12 entrance fee.

Set # 22, page 94

1. Slope $= -3$
   y-intercept $= -1$

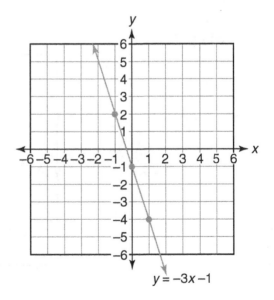

$$y = -3x - 1$$

2. Slope = $\dfrac{1}{2}$

   y-intercept = 2

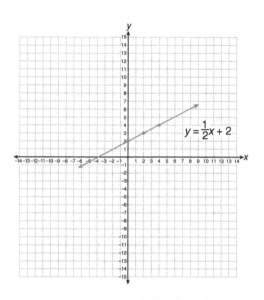

$$y = \frac{1}{2}x + 2$$

3. $y = -x - 5$

   Slope = -1

   y-intercept = -5

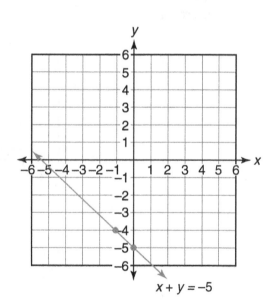

$$x + y = -5$$

4. $y = -x + 5$
   Slope $= -1$
   y-intercept $= 5$

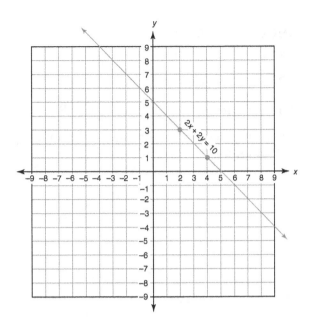

5. Slope $= 12$, $12 per day to rent y-intercept $= 40$, one-time fee of $40

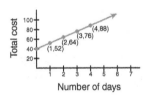

Set # 23, pages 99–100

1. Yes, 4/1

2. No

3. No

4. No

5. No

6. No

7. $(2, 10)$ and $(5, 25)$

8. $(8, 12)$ and $(10, 15)$

9. $y = 3x$, it passes through $(0, 0)$

Set # 24, page 103

1. x-intercept $= 10$, y-intercept $= 5$

2. x-intercept $= 5$, y-intercept $= -5$

3. x-intercept $= -4$, y-intercept $= 8$

4. x-intercept $= 6$, coordinates of x-intercept $= (6, 0)$
   y-intercept $= 3$, coordinates of y-intercept $= (0, 3)$

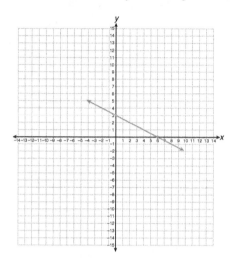

5. x-intercept $= -3$, coordinates of x-intercept $= (-3, 0)$
   y-intercept $= 3$, coordinates of y-intercept $= (0, 3)$

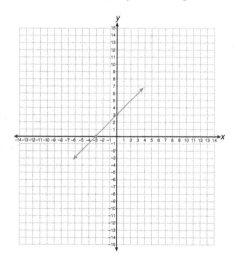

Set # 25, page 110

1. $y = 4x + 10$

2. $y = \dfrac{1}{2}x$

3. $y = -5x - 3$

4. $y = -\dfrac{2}{3}x + 4$

5. $y = 5$

6. $y = 3x + 26$

7. $y = 2x - 2$

8. $y = \dfrac{1}{3}x + 1$

9. $y = -4x + 18$

10. $y = -2x + 12$

11. $y = 2x - 4$

Set # 26, page 114

1. $y = 1.25x + 10.50$

2. a. $C = 33m + 10$

   b. $604

3. $y = -4x - 2$

4. $y = 0.45x - 0.25$

Chapter 4

Systems of Equations

Solve by Graphing

A **system of linear equations** consists of two or more lines. The solution to the system is the point(s) that the lines have in common. Three situations are possible.

The lines intersect at one point, having *one solution*.

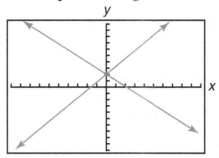

The lines are parallel and never intersect, therefore having *no solution*.

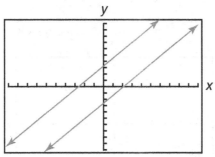

The lines are the same and intersect at every point. For this, the solution is *all real numbers* or *infinite solutions*.

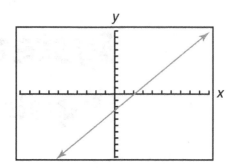

To solve a system of equations graphically:

Step 1: Graph the first equation.

Step 2: Graph the second equation on the same axes.

Step 3: Find the point where they intersect (solution).

Step 4: Check the point in both original equations.

Example 1:

Solve graphically $y = 2x + 1$ and $y = -\dfrac{1}{2}x + 6$.

Step 1: Graph the first line.

$y = 2x + 1$ $\rightarrow$ Slope $= 2$, y-intercept $= 1$

Step 2: Graph the second line.

$y = -\dfrac{1}{2}x + 6$ $\rightarrow$ Slope $= -\dfrac{1}{2}$, y-intercept $= 6$

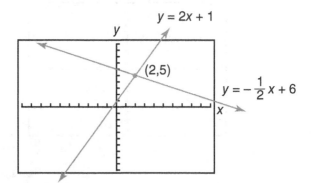

Step 3: Find the point where the graphs intersect. The solution for this system is $(2, 5)$.

Step 4: Substitute $(2, 5)$ into both equations to check.

$$y = 2x + 1 \qquad\qquad y = -\frac{1}{2}x + 6$$

$$5 = 2(2) + 1 \qquad\qquad 5 = -\frac{1}{2}(2) + 6$$

$$5 = 5 \text{ Check!} \qquad\qquad 5 = -1 + 6$$

$$5 = 5 \text{ Check!}$$

Example 2:

Let's try one more.

$$\begin{cases} y = -x + 4 \\ y = 2x + 1 \end{cases}$$

Graph the lines.

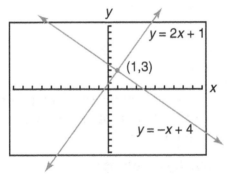

$$y = -x + 4 \qquad \rightarrow \quad \text{Slope} = -1, y\text{-intercept} = 4$$

$$y = 2x + 1 \qquad \rightarrow \quad \text{Slope} = 2, y\text{-intercept} = 1$$

Find the point of intersection $(1, 3)$.

Check: $\quad y = -x + 4 \qquad y = 2x + 1$

$\qquad\qquad 3 = -1 + 4 \qquad 3 = 2(1) + 1$

$\qquad\qquad 3 = 3 \text{ Check!} \quad 3 = 2 + 1$

$\qquad\qquad\qquad\qquad 3 = 3 \text{ Check!}$

Example 3—Solve by graphing:

Solve graphically: $y = 2x + 1$ and $y - 2x = 5$

$y = 2x + 1$	$y - 2x = 5$
Slope = 2	$y = 2x + 5$
y-intercept = 1	Slope = 2
	y-intercept = 5

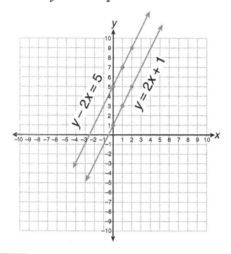

No Solution

Notice the slopes are the same; therefore, the lines are parallel. Parallel lines will never intersect; therefore, no common point of intersection.

Example 4—Solve by graphing:

Solve graphically: $y = -2x + 1$ and $y + 2x = 1$

$y = -2x + 1$	$y + 2x = 1$
Slope = -2	$y = -2x + 1$
y-intercept = 1	Slope = -2
	y-intercept = 1

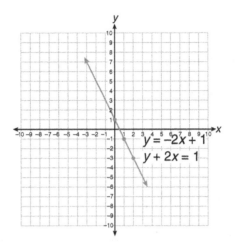

$y = -2x + 1$
$y + 2x = 1$

Infinite Solutions

Notice the lines are exactly the same. Therefore, any number will work in either equation.

Now **without graphing** (look at the equations, what do you know?) determine what type of solution each of the following systems will have. This is called solving by **inspection**.

1. $y = -2x + 5$
   $y = x + 4$

   Different slopes, **1 solution**.

2. $x + y = 1$
   $x + y = 3$

   Same slope, different y-intercepts
   Lines will be parallel; therefore, **no solution**.

3. $x + y = 5$
   $2y = -2x + 10$

   You might need to rearrange the equations before you notice any patterns.
   $x + y = 5$ is the same as $y = -x + 5$
   $2y = -2x + 10$ is the same as $y = -x + 5$

   Notice, now you can tell they are the same equation; therefore, **infinite solutions**.

BRAIN TICKLERS   Set # 27

Solve graphically and check.

1. $\begin{cases} y = -x + 5 \\ y = 2x - 1 \end{cases}$

2. $\begin{cases} y = 3x + 1 \\ x + y = 1 \end{cases}$

3. $\begin{cases} y = -x + 3 \\ y - 1 = x \end{cases}$

4. Graph the lines $y = 2x + 4$ and $y = 2x - 1$ on the same axes. Explain why there is no solution.

Without graphing, explain how many solutions each system has. Justify your reasoning.

5. $2x + 2y = 12$
$x + y = 6$

6. $x + y = 10$
$y = -x + 3$

7. $y = x + 1$
$2y = -6x - 8$

(Answers are on pages 141–142.)

Solve Using Substitution

You can also solve a system of equations algebraically, that is by using the equations and not graphing them. You have substituted before with problems such as this: Evaluate $2x + 5$, if $x = -2$.

To do it, you would substitute the value -2 for x and then simplify. So $2(-2) + 5 = 1$.

To solve a system of equations using substitution, substitute one equation into the other equation. You are trying to solve for only one variable at a time. Look at the following example to see how this works.

Example 1:

Solve algebraically: $y = x - 8$
$$y = 5x$$

Step 1: Both equations need to be in the form $y =$ (they both already are).
$$y = x - 8$$
$$y = 5x$$

Step 2: Since you know what $y =$, substitute for y in the second equation.

$y = x - 8$ Substitute for y. We know it is $x - 8$, so put this in place of the y in the second equation.

$x - 8 = 5x$

$x - 8 = 5x$ Now we have one variable so we can solve for x.

$-8 = 4x$
$x = -2$

Step 3: Since we know x, plug the x-value back into *one of the original equations* to find y. (It doesn't matter which equation you choose.)
$$y = 5x$$
$$y = 5(-2)$$
$$y = -10$$

Step 4: Write the solution as an ordered pair:
Solution: $(-2, -10)$

What does this mean? If you were to graph both lines, they would intersect at $(-2, -10)$.

We proved it algebraically. Here is the graph to also prove it.

$y = x - 8$　　　　　　$y = 5x$
Slope $= 1$　　　　　　Slope $= 5$
y-intercept $= -8$　　　y-intercept $= 0$

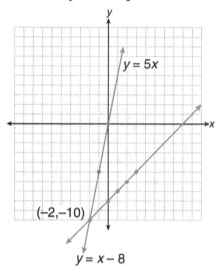

Solution: $(-2, -10)$

Example 2:

Solve algebraically:　$y = x$
　　　　　　　　　　$2x + y = 3$

Step 1: Both equations need to be in the form
　　　　　$y =$ (we have one already).

　　　　　　　　$y = x$
　　　　　　　　$2x + y = 3$　⟶　needs to be
　　　　　　　　　　　　　　　　　$y = -2x + 3$

Step 2: Since you know what $y =$, substitute for y in the
　　　　　second equation.

　　　　$y = x$　　　⟩　Since both equations are now in
　　　　　　　　　　　　the form $y =$, set them equal to
　　　　　　　　　　　　each other.

　　　　$y = -2x + 3$
　　　　$x = -2x + 3$　　Now we have one variable so we
　　　　　　　　　　　　can solve for x.

　　　　$3x = 3$
　　　　$x = 1$

Step 3: Since we know x, plug the x-value back into *one of the original equations* to find y.

$$y = x$$
$$y = 1$$

Step 4: Write the solution as an ordered pair:
Solution: $(1, 1)$

What does this mean? If you were to graph both lines, they would intersect at $(1, 1)$.

We proved it algebraically. Here it is graphically.

$y = x$	$2x + y = 3$
Slope $= 1$	$y = -2x + 3$
y-intercept $= 0$	Slope $= -2$
	y-intercept $= 3$

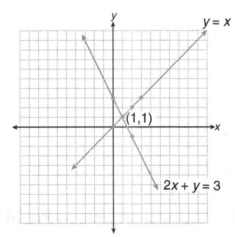

Solution: $(1, 1)$

To solve a system of equations algebraically:

- Write both equations in the form $y = mx + b$.
- Set the two equations equal to each other.
- Solve for x.
- Plug into an original equation to solve for y.
- Write your solution as an ordered pair.

BRAIN TICKLERS   Set # 28

Solve the following systems of equations algebraically. Check.

1. $y = x + 2$
$y = -3x$

2. $y = -x - 4$
$y = x$

3. $y = x + 1$
$x + y = 9$

4. $y = 2x$
$x + y = 21$

(Answers are on page 143.)

Systems of Equations Word Problems

Use what you have learned about systems of equations to practice the following. Try to set up the problems on your own, and then check with the solutions to the following examples to see if you are doing it correctly.

PAINLESS TIP

When writing equations remember to:

- Define your variables.
- Write your two equations (your system).
- Solve for $y =$ if necessary.
- Set them equal to each other and solve for one letter.
- Substitute your first answer into an original equation to get the second value.
- Write your solution as a coordinate (x, y).
- Check.

Examples:

Solve each system graphically and algebraically.

1. y equals three more than negative four times a number x. The sum of negative x and y equals negative two. Find the numbers.

   Solution:

   Variables: Let $x =$ a number
   Let $y =$ a number

   Equations: $y = -4x + 3$
   $-x + y = -2$

   Graphically:

   $y = -4x + 3$ $-x + y = -2$
   Slope $= -4$ $y = x - 2$
   y-intercept $= 3$ Slope $= 1$
   y-intercept $= -2$

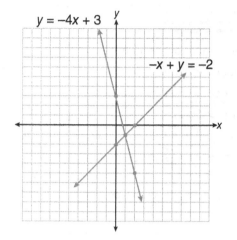

Solution: $(1, -1)$
The numbers are 1 and -1.

Algebraically:

$$y = -4x + 3$$
$$-x + y = -2 \qquad \text{(Need this one to be in form } y=.\text{)}$$

$$y = -4x + 3$$
$$\underline{y = x - 2}$$

$-4x + 3 = x - 2$	$y = -4x + 3$
$\underline{-x \qquad\quad -x}$	$y = -4(1) + 3$
$-5x + 3 = -2$	$y = -1$
$-5x = -5$	
$x = 1$	

Solution: $(1, -1)$
The numbers are 1 and -1.

2. A rectangle has a perimeter of 18 centimeters. Its length is 5 centimeters longer than its width. Find the dimensions of the rectangle.

Solution:

Variables: Let $x =$ width
               Let $y =$ length

System: $y = 5 + x$
                         $2x + 2y = 18$

Graphically:

$y = 5 + x$	$2x + 2y = 18$
Slope $= 1$	$x + y = 9$
y-intercept $= 5$	$y = -x + 9$
	Slope $= -1$
	y-intercept $= 9$

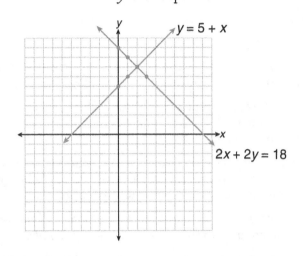

Solution: $(2, 7)$

The rectangle has a width of 2 centimeters and a length of 7 centimeters.

Algebraically:

$$y = 5 + x$$
$$2x + 2y = 18 \qquad \text{(Need this one to be in form } y =.)$$

$$y = 5 + x$$
$$\underline{y = -x + 9}$$

$$
\begin{array}{ll}
5 + x = -x + 9 & y = 5 + x \\
\underline{\quad +x \qquad\quad +x} & y = 5 + 2 \\
5 + 2x = 9 & y = 7 \\
\underline{-5 \qquad\quad -5} & \\
\dfrac{2x}{2} = \dfrac{4}{2} & \\
x = 2 &
\end{array}
$$

Solution: $(2, 7)$

The rectangle has a width of 2 centimeters and a length of 7 centimeters.

BRAIN TICKLERS   Set # 29

Write a system of equations that represents each situation.

- Define your variables—state what *x* represents and what *y* represents.

- Write the two equations to represent the situation.

- Solve algebraically.

1. Brenda and 23 of her friends went on a limo ride. There were 8 more boys than girls on the ride. How many boys and girls were on the ride?

2. Frank drove a total of 248 miles on Monday. He drove 70 fewer miles in the morning than he did in the afternoon. How many miles did he drive in the afternoon?

3. Chris is a cross-country ski racer. On Saturday, he skied twice as many miles as he did on Sunday. Over the weekend, he skied a total of 63 miles. How far did he ski on each day?

4. Mr. Stahl downloaded 34 more songs than Mr. Johnston. Together they downloaded 220 songs. How many songs did each person download?

5. The seventh- and eighth-grade bands held a joint concert. Together there were 188 band members. If the eighth-grade band is three times as big as the seventh-grade band, how big is the eighth-grade band?

(Answers are on page 143.)

BRAIN TICKLERS—THE ANSWERS

Set # 27, page 132

1. $(2, 3)$

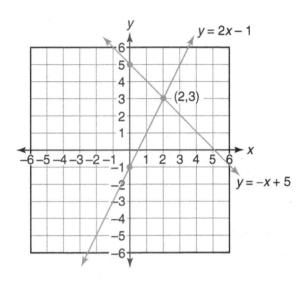

2. $(0, 1)$

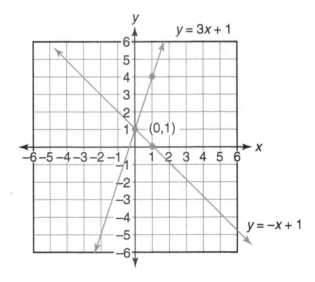

3. $(1, 2)$

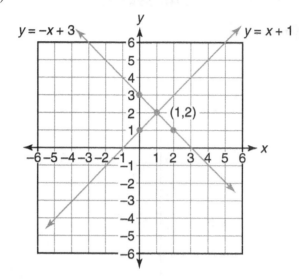

4. The lines are parallel. Since they never intersect, they have no solution.

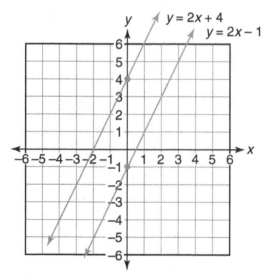

5. Infinite solutions. They are the same line.

6. No solution. The lines have the same slope (2), so they are parallel.

7. One solution. Different slopes, lines will cross once.

Set # 28, page 136

1. $(-0.5, 1.5)$
2. $(-2, -2)$
3. $(4, 5)$
4. $(7, 14)$

Set # 29, page 140

1. There were 16 boys and 8 girls.
2. He drove 89 miles in the morning and 159 miles in the afternoon.
3. He skied 42 miles on Saturday and 21 miles on Sunday.
4. Mr. Stahl downloaded 127 songs, and Mr. Johnston downloaded 93.
5. There are 47 seventh graders and 141 eighth graders.

Chapter 5

Introduction to Functions

Relations and Functions

A **relation** is a set of ordered pairs. For example: $\{(1,2), (3,4), (5,6)\}$.

In a relation, the first element is called the **domain**. The second element is called the **range**.

Want to make this *painless?*

- Domain is the set of x-values.
- Range is the set of y-values.

When looking at a list of ordered pairs, we used these symbols $\{\quad\}$ to write a set.

Look at these relations (set of ordered pairs):

$(2,3)\ (4,5)\ (6,7)$

Domain: $\{2,4,6\}$
Range: $\{3,5,7\}$

x	y
2	4
6	8
10	12
14	16

Domain: $\{2, 6, 10, 14\}$
Range: $\{4, 8, 12, 16\}$

A **function** is a relation in which the element of the domain corresponds to one and only one element in the range.

What is an easy way to say this? Every x-value can have only one y-value.

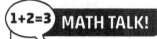

Relation: a set of ordered pairs

Domain: the set of all *x*-values

Range: the set of all *y*-values

Function: every *x*-value has only one *y*-value

Relations that
ARE functions:

$(2, 5)$ $(4, 7)$ $(10, 6)$

$$2 \longrightarrow 5$$
$$4 \longrightarrow 7$$
$$10 \longrightarrow 6$$

Domain: $\{2, 4, 10\}$

Range: $\{5, 7, 6\}$

A function because every
x-value has only one
y-value.

Relations that
are NOT functions:

$(1, 2)$ $(1, 3)$ $(2, 5)$ $(3, 7)$

$$1 \longrightarrow 2$$
$$2 \longrightarrow 3$$
$$3 \longrightarrow 5$$
$$7$$

Domain: $\{1, 2, 3\}$
(It is not necessary to
write 1 twice.)
Range: $\{2, 3, 5, 7\}$

Not a function because the
number 1 has two different
y-values.

Functions can be represented in different ways, including tables, graphs, and equations. If you can plug values into an equation/ function, then, remember, you can graph the line! And we will prove it's a function.

Example 1:

Graph the line $y = 2x + 5$. Determine if it is a function.

Solution:

$$\text{Slope} = \frac{2}{1}$$

y-intercept $= 5$

Is it a function?
Yes, every x-value has only one y-value.

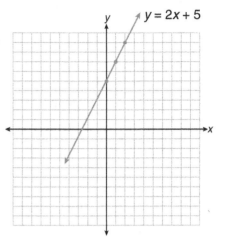

How can we prove this is a function? Looking at the graph, you can use the **vertical line test**.

Vertical line test

Place a vertical line anywhere on the graph. If the line intersects (crosses) the graph *only once*, then it is a function.

Examples:

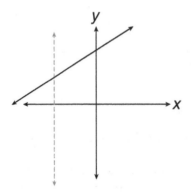

Function—crosses once

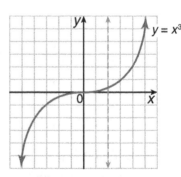

Function—crosses once

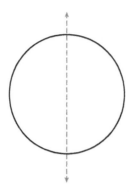

NOT a function—crosses twice!

Example 2:

Graph $y = x^2$ for the values $-3 \leq x \leq 3$. Then determine if this is a function.

Solution:
Make a table of values and graph.

x	x²	y
−3	$(-3)^2$	9
−2	$(-2)^2$	4
−1	$(-1)^2$	1
0	0^2	0
1	1^2	1
2	2^2	4
3	3^2	9

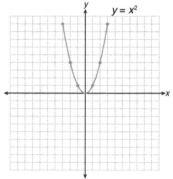

Is it a function? Yes, every *x*-value has only one *y*-value.

Notice this is a function—just not a *linear* function (it does not make a straight line). There are other functions that exist besides lines. You would call this a **nonlinear** function.

BRAIN TICKLERS   Set # 30

Find the domain and range of each relation. Then determine if each relation is a function. Explain your reasoning.

1.

Input	Output
−1	2
0	4
1	6
2	8

2. (0, 1) (1, 2) (2, 3) (3, 4)

3.

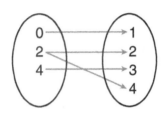

4. Given the relation:

$\{(1, 5) (6, 3) (8, 12) (x, 15) (−2, −3)\}$

- State a value of *x* that makes it a function.
- State a value for *x* that will not make it a function.

5. Determine if these are functions using the vertical line test, and explain your reasoning.

a.

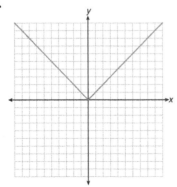

b.

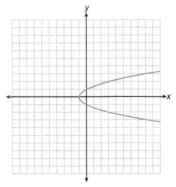

6. Graph the equation $y = \dfrac{1}{2}x + 3$. Is this a function?

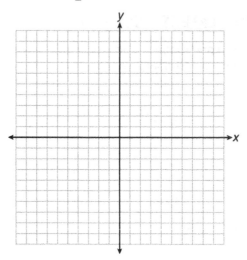

7. Given the following data:

x	0	2	4	6	2
y	1	3	6	−2	5

   a. What is the domain? What is the range?

   b. Does the relationship represent a function?

   c. Graph the data to verify your answer in part b.

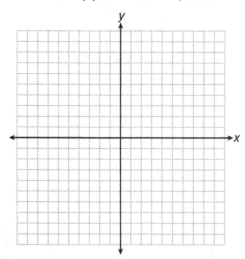

(Answers are on pages 166–167.)

Function Notation

Functions can be represented in different ways, including tables, equations, and graphs. You have actually worked with functions in this book so far, but function notation has not been formally introduced until now.

Old way: Function notation way:

$$\frac{y = mx + b}{y = 2x + 5} \qquad f(x) = 2x + 5$$

All we did was replace y with $f(x)$, which is read "f of x."

Example:

1. Write $y = -3x$ using function notation.

Solution: $f(x) = -3x$
Even though it is common to use the letter f when writing function notation, any letter can be used, either lowercase or capital.

$$f(x) = 4x \qquad g(x) = -x - 1 \qquad h(x) = 5 - 2x$$

Evaluating functions

You have been evaluating expressions for a long time. This is what you are familiar with:

Evaluate $y = 2x + 5$ for $x = 3$.

The same problem can be written in function notation:

Find $f(3)$ when $f(x) = 2x + 5$.

} Same problem, **different notation**

Instead of $y = 2x + 5$, we say $f(x) = 2x + 5$, read as "f of x equals $2x + 5$."

How would you evaluate a problem written in function notation?

Solution:

Old way:

Evaluate $y = 2x + 5$ when $x = 3$.

$y = 2(3) + 5$

$y = 6 + 5$

$y = 11$

Function notation way:

Find $f(3)$ when $f(x) = 2x + 5$.

This notation is telling you to plug 3 into the f function for x.

$f(3) = 2(3) + 5$

$f(3) = 6 + 5$

$f(3) = 11$

↑

This means when $x = 3$, $y = 11$.

Here is another problem, same problem, worded two ways:

Old way:

Evaluate $y = -4x$ when $x = 2$.

$y = -4(2)$

$y = -8$

Function notation way:

Find $f(2)$ when $f(x) = -4x$.

$f(2) = -4(2)$

$f(2) = -8$

↑

This means $y = -8$.

More examples:

2. Evaluate the following, given the three functions.

$$f(x) = 2x - 1 \qquad g(x) = -2x \qquad h(x) = 4 - x$$

a. Find $f(-2)$.

↑

This means you are subbing -2 into the f function.

Old wording:

Evaluate $y = 2x - 1$ when $x = -2$.

$y = 2(-2) - 1$

$y = -5$

New wording:

Given $f(x) = 2x - 1$, find $f(-2)$.

$f(-2) = 2(-2) - 1$

$f(-2) = -4 - 1$

$f(-2) = -5$

So when $x = -2, y = -5$.   So when $x = -2, y = -5$.

Notice you get the same answer! They are the same problem, "function" just makes it sound fancier!

Evaluate the following, given the three functions.

$$f(x) = 2x - 1 \qquad g(x) = -2x \qquad h(x) = 4 - x$$

b. Find $h(5)$.

This means you are subbing 5 into the h function.

Old wording:	New wording:
Evaluate $y = 4 - x$ when $x = 5$.	Given $h(x) = 4 - x$, find $h(5)$.
$y = 4 - (5)$	$h(5) = 4 - 5$
$y = -1$	$h(5) = -1$
So when $x = 5, y = -1$.	So when $x = 5, y = -1$.

c. Find $g(-4)$.

Pick the $g(x)$ function!

$g(x) = -2x$

$g(-4) = -2(-4)$

$g(-4) = 8$

input/domain   output/range
x y

d. Find when $f(x) = 0$.

This time you want the f function set equal to 0.

$$2x - 1 = 0$$
$$\underline{ +1 +1}$$
$$2x = 1$$

$$x = \frac{1}{2}$$

So $f(x) = 0$, when $x = \frac{1}{2}$.

1+2=3 MATH TALK!

Recall:

Domain—the set of all x-values

Range—the set of all y-values

Function—every x-value has only one y-value

BRAIN TICKLERS   Set # 31

1. Given the function $f(x) = 2x - 5$, find $f(-2)$.

2. Find $g(12)$ if $g(x) = 0.5x + 7$.

3. Given the functions:

$$f(x) = x^2$$
$$g(x) = -3x - 1$$
$$h(x) = -2 - x$$

Evaluate:

a. $g(3)$

b. $f(-4)$

c. $h(-8)$

d. $f(2) + g(1)$

e. $g(0) + h(6)$

f. $f(1) + g(2) + h(3)$

g. $h(x) = 0$

h. $2(f(4))$

i. $3(f(x))$

4. Evaluate the equation $h(x) = -3x + 1$ using the values $-1 \le x \le 2$, and determine if this is a function.

(Answers are on page 167.)

Evaluating Functions from a Graph

You know now how to evaluate a function given the equation. You can also evaluate functions given the graphs.

Remember, when in the form $y = mx + b$ or $f(x) = mx + b$, this is a linear function. We can graph a function using a table of values or slope and y-intercept.

Example 1:

Graph the function $f(x) = -x - 1$ and evaluate $f(2)$.

Solution:
Make a table

x	$-x - 1$	y
-2	$-(-2) - 1$	1
0	$0 - 1$	-1
2	$-2 - 1$	-3

or

Use slope-intercept

$$\text{Slope} = \frac{-1}{1}$$

$$y\text{-intercept} = -1$$

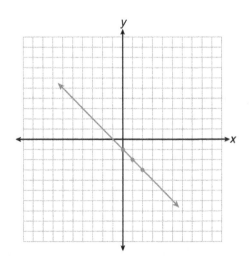

If you want to find $f(2)$, you can do this two ways:

- Look at the table when $x = 2$, and see that $y = -3$.
- Look at the graph when $x = 2$, and see the coordinate $(2, -3)$, so $y = -3$.

Example 2:

Using the graph, find the following:

Find $f(0)$.
$f(0) = 3$, because when $x = 0, y = 3$.

Find $f(3)$.
$f(3) = -1$, because when $x = 3, y = -1$.

Find $f(x) = -2$.
$x = -4$, because when $x = -4, y = -2$.

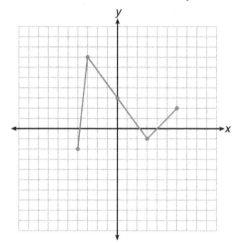

Example 3:

Given three functions:

$h(x) = 3x^2 - 2x$

x	f(x)
22	9
4	6
6	3

Evaluate:

a. $f(4)$
   $f(4) = 6$

b. $g(3)$
   $g(3) = 0$

c. $h(-2)$
   $h(-2) = 3(-2)^2 - 2(-2)$
   $\quad\quad = 3(4) + 4$
   $\quad\quad = 12 + 4$
   $\quad\quad = 16$

d. $g(0) + f(6)$
   $g(0) = 5$
   $f(6) = 3$
   $5 + 3 = 8$

e. $g(x) = -3$
   when $x = -5, y = -3$

f. $h(m)$
   $h(m) = 3m^2 - 2m$

Example 4:

The figure below shows the graph of temperatures (in degrees Fahrenheit) during a 12-month period in Central Park.

a. If $T = m(x)$, find the value of $m(4)$.

When $x = 4$ (representing April), the temperature is about 55°F.

b. If you want to tour Central Park when it is above 80°F, when should you plan your visit?

July, August, September because the temperature is more than 80°F.

c. Estimate $m(11)$ and explain what it means in the context of the problem.

60°. It is about 60° in November in Central Park.

BRAIN TICKLERS Set # 32

Use the graph to find the following:

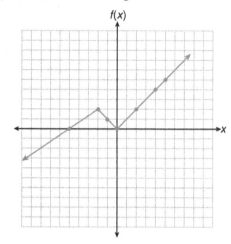

f(x)

1. $f(-1)$ **2.** $f(5)$ **3.** $f(x) = 4$

4. $h(x) = 0.5x + 4$

x	f(x)
0	−5
3	−1
6	3
9	7

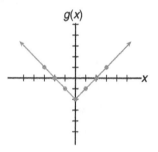

g(x)

Evaluate:

a. $f(3)$

b. $g(0)$

c. $h(10)$

d. $g(3) + h(-6)$

e. $f(x) = 7$

f. $(h(-20))^2$

5. Use the graph to answer the following questions:

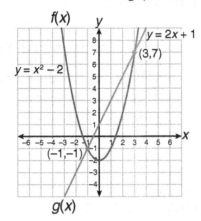

a. $f(0)$

b. $g(0)$

c. $f(3) + g(1)$

d. $f(x) = 2$

e. $g(x) = 5$

(Answers are on page 168.)

Comparing Functions

Which function has the greatest rate of change? The smallest rate of change? Order the slopes from least to greatest. Yikes!

Let's make this *painless!* Comparing functions is basically a huge slope review! Up until this point, you have used a lot of different skills, but in these questions, you need to think of everything you know about lines/functions and slope.

Here is a reminder:

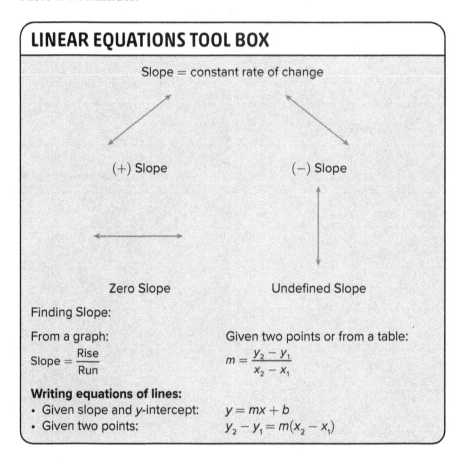

LINEAR EQUATIONS TOOL BOX

Slope = constant rate of change

(+) Slope

(−) Slope

Zero Slope

Undefined Slope

Finding Slope:

From a graph:

$$\text{Slope} = \frac{\text{Rise}}{\text{Run}}$$

Given two points or from a table:

$$m = \frac{y_2 - y_1}{x_2 - x_1}$$

Writing equations of lines:
- Given slope and y-intercept: $y = mx + b$
- Given two points: $y_2 - y_1 = m(x_2 - x_1)$

Let's go!

Example 1:

Given the three functions:

$$f(x) \qquad g(x) = x + 5 \qquad h(x)$$

x	f(x)
3	6
4	8
5	10

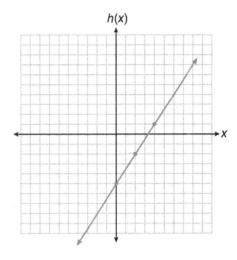

Which function has the greatest rate of change? Justify your answer.

Solution:
In order to do this, you have to find the rate of change/slope for each function.

$$f(x) = \frac{\text{change in } y}{\text{change in } x} = \frac{2}{1} \qquad g(x) = 1 \qquad h(x) = \frac{3}{2}$$

$f(x)$ has the greatest rate of change since 2 is greater than 1 and $\frac{3}{2}$. Remember $\frac{3}{2} = 1.5$.

Example 2:

Which function has the greatest rate of change?

(a)   (b)   (c)

$y = 6 - 2x$

x	y
2	−12
0	0
−2	12
−4	24

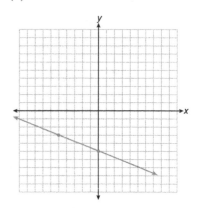

Solution:
Find the rate of change for each example.

a. Slope $= \dfrac{-2}{1}$

b. $(2, -12)$ and $(0, 0)$

$$\frac{0 - (-12)}{0 - 2} = \frac{12}{-2} = -6$$

c. $\dfrac{\text{Rise}}{\text{Run}} = \dfrac{2}{-5}$

Choice (b) has the greatest rate of change. Remember the negative tells us the "direction" of the slope of the line, the rate of change is 6, which is greater than 2 and $\dfrac{2}{5}$.

Example 3:

The total cost of renting a carpet cleaner from EZ Clean is represented by the function $y = 25x + 15$, where x represents the number of days and y represents the total cost. The cost of renting a carpet cleaner from Radiant Rugs is represented in the table below:

Number of days (x)	Total cost (y)
1	45
2	60
3	75
4	90
5	105

a. State and interpret the rate of change for each business.

EZ Clean $= 25/1$, which means it costs $25 per day to rent the machine.

Radiant Rugs $= 15/1$, which means it costs $15 per day to rent the machine.

b. Find and interpret the initial fee for both companies.

EZ Clean charges a $15 fee, and Radiant Rugs charges $30 (find when $x = 0$).

c. Which company has the greater rate of change?

EZ Clean has a higher rate of change, which is $25 per day.

d. Which company should you use if you want to rent the carpet cleaner for 8 days? Justify your answer.

EZ Clean for 8 days $= 25(8) + 15 = 215.

Radiant Rugs $=$ continue chart $=$

6	120
7	135
8	150

Radiant Rugs is better; it costs only $150 compared to $215.

e. Are these both functions? Explain your reasoning.

Yes, because for both, every x-value has only one y-value.

BRAIN TICKLERS   Set # 33

1. Three sisters are comparing their phone plans. Amy's monthly fee is represented by the function $A(x) = 0.05x + 20$, in which x represents the number of minutes. Tracy's plan is represented by the table, and Kelly's plan is represented by the graph.

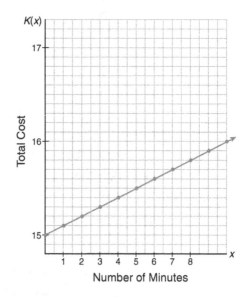

x	T(x)
0	0
1	0.15
2	0.30
3	0.45

a. Write an equation for Tracy's and Kelly's plans.

b. Whose plan has the highest rate per minute? Why?

c. How do the initial fees compare for all three plans?

d. When will Amy's and Tracy's plans cost the same?

e. Choosing between all three plans, which plan(s) is best if you talk for 300 minutes?

2. Melissa loves to read. While reading her last three books, she kept track of how long it took her to read each book. Her data is shown below:

Book A:

# of Days	0	1	2	3	4
Pages Left	600	550	500	450	400

Book B: Has 400 pages, and she reads 25 pages a day.

Book C: $y = 388 - 36x$, where x represents number of days.

Which book took Melissa the longest to read? Show all your work and justify your answer.

3. Owen just graduated from college and has been given two different job offers. For Job A, Owen would make a starting salary of $20,000 with a yearly increase of $2,500. For Job B, he would make a starting salary of $25,000 and get a $2,000 yearly increase.

a. If Owen were to work at this new job for only 5 years, which job should he take and why?

b. You would like to offer Owen a job with your company. Here is your offer:

0	1	2	3
$18,000	$21,200	$24,400	$27,600

What job should Owen choose if he wants to work for 10 years? Explain.

4. Function A has a slope of 0.75 and a y-intercept of 2.

a. Write the equation of a line with a greater rate of change and the same y-intercept.

b. Write the equation of a line with a smaller rate of change and a greater y-intercept.

c. Write the equation of a line parallel to the given function.

(Answers are on page 168.)

BRAIN TICKLERS—THE ANSWERS

Set # 30, pages 149-150

1. D: $\{-1, 0, 1, 2\}$
   R: $\{2, 4, 6, 8\}$
   Yes, every x-value has only one y-value.

2. D: $\{0, 1, 2, 3\}$
   R: $\{1, 2, 3, 4\}$
   Yes, every x-value has only one y-value.

3. D: $\{0, 2, 4\}$
   R: $\{1, 2, 3, 4\}$
   No, 2 has more than one y-value.

4. $x =$ any number other than 1, 6, 8, or -2.
   If $x = 1, 6, 8,$ or -2, it will not be a function.

5. a. Yes, every x-value has only one y-value.

   b. No, the line crosses in more than one point.
      Therefore, there are repeated x-values.

6. Slope $= \dfrac{1}{2}$, y-intercept $= 3$. Yes, it is a function.

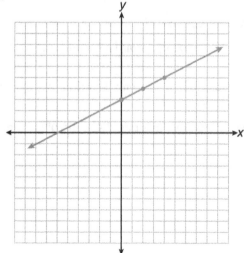

7. a. D: $\{0, 2, 4, 6\}$
     R: $\{1, 3, 6, -2, 5\}$

   b. No, there are repeated x-values.

   c. See graph.

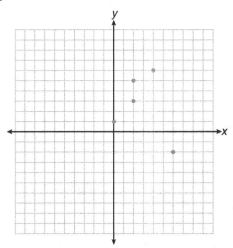

Set # 31, page 154

1. -9

2. 13

3. a. -10 d. 0   g. -2
   b. 16   e. -9 h. 32
   c. 6   f. -11 i. $3x^2$

4.

x	$-3x + 1$	y
-1	$-3(-1) + 1$	4
0	$-3(0) + 1$	1
1	$-3(1) + 1$	-2
2	$-3(2) + 1$	-5

Yes, it is a function; every x-value has only one y-value.

Set # 32, pages 158–159

1. 1

2. 5

3. $x = 4$

4. a. -1 c. 9   e. 9
   b. -2 d. 2   f. 36

5. a. -2 c. 10   e. $x = 2$
   b. 1   d. $x = 2$ and -2

Set # 33, pages 164–165

1. a. Tracy $= 0.15x$, Kelly $= 0.10x + 15$
   b. Tracy's, it's 0.15/min.
   c. Amy's $= \$20$, Tracy's $= \$0$, Kelly's $= \$15$
   d. 200 minutes
   e. Amy's, because hers will only cost $35, and Tracy's and Kelly's will each cost $45.

2. Book A: $y = 600 - 50x$, took 12 days
   Book B: $y = 400 - 25x$, took 16 days
   Book C: $y = 388 - 36x$, took about 11 days
   So Book B took the longest to read.

3. a. Job A: $20,000 + 2,500x$
   Job B: $25,000 + 2,000x$
   He should choose Job B because he will make $32,500 at Job A and $35,000 at Job B.
   b. After 10 years, Job A would pay $45,000, Job B would pay $45,000, and your company would pay $50,000. He should take the job offer from you.

4. a. $y = 2x + 2$ (the slope can be any number greater than 0.75)
   b. $y = 0.3x + 8$ (the slope can be any number smaller than 0.75, and the y-intercept larger than 2)
   c. $y = 0.75x$ (any line with the same slope and any y-intercept)

Chapter 6

Triangles and Angles

Types of Angles

There are many different types of angles. Here is a summary of angle vocabulary you have probably heard in math before.

Complementary angles

Two angles that add up to 90 degrees

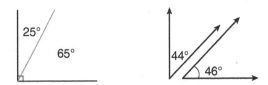

Supplementary angles (linear pair)

Two angles that add up to 180 degrees

Vertical angles

Two angles that are formed by intersecting lines
Vertical angles are equal (congruent)

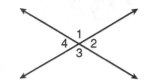

measure of ∠1 = measure of ∠3
measure of ∠2 = measure of ∠4

Symbol for congruent: ≅
∠1 ≅ ∠3 and ∠2 ≅ ∠4

Adjacent angles

Two angles that have a common vertex and share a common side

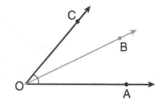

∠*COB* is adjacent to ∠*BOA*.

1+2=3 MATH TALK!

When describing the measurement of angles, **complementary**,
supplementary, and **vertical** are used. **Adjacent** is just a description of
location, but the other three explain the type of angle and, therefore,
the relationship of the angle measures.

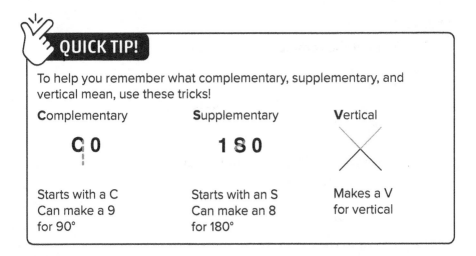

QUICK TIP!

To help you remember what complementary, supplementary, and vertical mean, use these tricks!

Complementary	**S**upplementary	**V**ertical
C 0	**1 S 0**	
Starts with a C Can make a 9 for 90°	Starts with an S Can make an 8 for 180°	Makes a V for vertical

Angles in shapes

The angles in a triangle add up to 180 degrees.

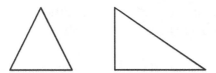

The angles in a quadrilateral add up to 360 degrees.

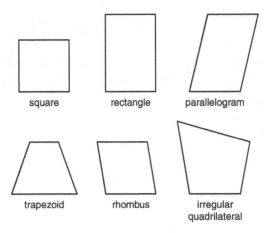

square   rectangle   parallelogram

trapezoid   rhombus   irregular quadrilateral

Examples:

1. Find the complement of each angle.

   a. 23°   b. 61°

   Solution: Solution:
   $90 - 23 = 67°$ $90 - 61 = 29°$

2. If $m\angle 1 = 47°$ and $\angle 1$ and $\angle 2$ are complementary angles, find $m\angle 2$.

   Solution:
   $$47 + x = \quad 90$$
   $$\underline{-47 \qquad -47}$$
   $$x = \quad 43 \qquad \text{The measure of } \angle 2 = 43°.$$

3. Find the supplement of each angle.

   a. 164°   b. 75°

   Solution: Solution:
   $180 - 164 = 16°$ $180 - 75 = 105°$

4. If $m\angle 2 = 53°$ and $\angle 1$ and $\angle 2$ form a linear pair, find $m\angle 1$.

   Solution:
   $$53 + x = 180$$
   $$\underline{-53 \qquad -53}$$
   $$x = 127 \qquad \text{The measure of } \angle 1 = 127°.$$

5. Use the diagram below. List two sets of vertical angles and four sets of supplementary angles. Explain your reasoning.

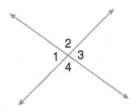

Solution:

∠1 and ∠3 are vertical angles. ∠2 and ∠4 are also vertical angles. Vertical angles are opposite each other.

∠1 + ∠2 = 180, ∠2 + ∠3 = 180, ∠3 + ∠4 = 180 and ∠1 + ∠4 = 180. Supplementary angles are two angles that add up to 180°.

6. In the diagram below, line PQ intersects line RT at point S, and the measure of $\angle TSQ$ is 88°.

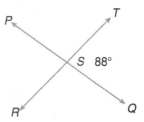

a. Find the measure of $\angle PST$. Show your work.

Solution:
180 − 88 = 92°

b. Name an angle vertical to $\angle TSQ$. Find its measure.

Solution:
$\angle PSR$ is vertical. $\angle PSR$ = 88°.

7. Two angles in a triangle add up to 136°. What is the measure of the third angle?

Solution:
136 + x = 180, so x = 44°

8. Two angles in a triangle are equal. If the third angle is 76°, find the measures of the other two angles.

Solution:

$x + x + 76 = 180$ $180 - 76 = 104$ and $104/2 = 52°$

$2x + 76 = 180$

$\underline{-76 \quad -76}$

$2x \qquad = 104$

$x \qquad = 52°$

9. In the diagram below, triangle BAD is a right triangle. Find the measure of $\angle BAC$.

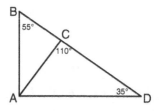

Solution:

Find $\angle CAD$:

$110 + 35 = 145$ and $180 - 145 = 35°$

So $\angle CAD = 35°$

Since $\angle BAD$ is a right angle, $35 + x = 90$, so $\angle BAC = 55°$.

BRAIN TICKLERS   Set # 34

1. Find the complement of a 44° angle.

2. Find the supplement of a 63° angle.

3. What is the measure of the complement of an angle represented by $5x$?

4. If one angle measures $7x$, what is its supplement?

5. $\angle ACB$ and $\angle DCF$ are vertical angles. If $\angle ACB$ measures 46°, what is the measure of $\angle DCF$?

6. What is the measure of ∠x in the diagram below? Justify your answer.

7. In the diagram below, line *AB* and line *CD* intersect at point *P*.

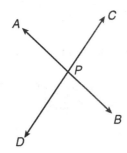

    a. Name an angle that is always congruent to ∠*APD*. How can you classify these two angles?

    b. Name an angle that is supplementary to ∠*APD*. Justify your answer.

(Answers are on page 201.)

Finding Measures of Angles Algebraically

Based on what you know about angles and solving equations, you can now solve for angles algebraically!

THINGS TO REMEMBER:

Complementary angles:
Two angles that add up to 90°.   _____ + _____ = 90

Supplementary angles:
Two angles that add up to 180°.   _____ + _____ = 180

Vertical angles:
Vertical angles are congruent.   _____ = _____

Examples:

1. Two angles are complementary. If the measure of one angle is twice the measure of the other, find the measures of each angle.

   Solution:
   $$x + 2x = 90$$
   $$3x = 90$$
   $$x = 30$$
   The measures of the angles are 30° and 60°.

2. A pair of supplementary angles are in the ratio 7:8. Find the measures of both angles.

   Solution:
   $$7x + 8x = 180$$
   $$15x = 180$$
   $$x = 12 \qquad 7(12) = 84 \text{ and } 8(12) = 96$$
   The measures of the angles are 84° and 96°.

3. Find the measures of the angles.

   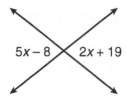

   Solution:
   $5x - 8 = 2x + 19$ since these are vertical angles.
   $$\begin{array}{r} 5x - 8 = 2x + 19 \\ \underline{-2x \qquad -2x} \\ 3x - 8 = 19 \\ \underline{+8 \quad +8} \\ 3x \qquad = 27 \\ x = 9 \qquad 5(9) - 8 = 37° \text{ and } 2(9) + 19 = 37° \end{array}$$

4. The supplement of an angle is 30° more than twice the angle. Find the angle and its supplement.

Solution:
Let x = one angle
Let $30 + 2x$ = the supplement of the angle

$$x + 30 + 2x = 180$$
$$3x + 30 = 180$$
$$\underline{-30 \quad -30}$$
$$3x = 150$$
$$x = 50$$

One angle is 50° and the other is $30 + 2(50) = 130°$.

BRAIN TICKLERS   Set # 35

1. If one angle is represented by the expression $(4x - 7)$, and its supplement is represented by the expression $(9x + 18)$, find the measures of these angles.

2. Two complementary angles are in the ratio of 5:13. Find the measures of each angle.

3. Two angles are supplementary. The larger angle is eight times greater than the smaller angle. Find the measures of the two angles.

4. One angle is 6 degrees more than twice the measure of its supplement. Find the measures of the angles.

5. Two angles are complementary. One angle is represented by $8x + 6$, and the other is represented by $19x + 3$. Find the measure of the larger angle.

(Answers are on page 201.)

The Sum of Angles in Triangles

The three angles inside a triangle add up to 180°. Do you know why?

Look at the following triangle. There are three angles in a triangle. Angles A, B, and C are known as the **interior angles** of the triangle. They are the angles "inside" the triangle. Let's lay angles A, B, and C next to each other.

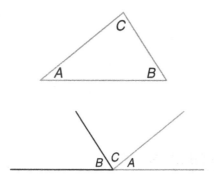

Notice that $\angle A + \angle B + \angle C$ make a straight line, and we know that a straight line measures 180°!

Examples:

1. Find the missing angle in the triangle.

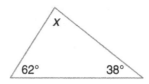

Solution:
$$62 + 38 + x = 180$$
$$100 + x = 180$$
$$-100 \qquad -100$$
$$x = 80$$

| The missing angle is 80°. |

2. In a triangle, the first angle measures three times the second, and the third angle measures 20° less than the second. Find the measure of each angle.

Solution:
Write an expression for each angle:
Let x = measure of second angle
Let $3x$ = measure of first angle
Let $x - 20$ = measure of third angle

Since angles in a triangle add up to 180°, write an equation and solve!

$$x + 3x + x - 20 = 180$$
$$5x - 20 = 180$$
$$\underline{+20 \quad +20}$$
$$\frac{5x}{5} = \frac{200}{5}$$
$$x = 40$$

Substitute 40 for x to find all three angles.

Let x = measure of second angle $= 40°$
Let $3x$ = measure of first angle $= 3(40) = 120°$
Let $x - 20$ = measure of third angle $= 40 - 20 = 20°$

| The angles of the triangle are 120°, 40°, and 20°. |

BRAIN TICKLERS   Set # 36

1. Find the measure of the base angles in an isosceles triangle if the third angle measures 47°.

2. A scalene triangle has angles that measure 8x, 6x, and 4x. Find the measures of the angles in the triangle.

3. A right triangle has one angle that measures 52°. Find the measure of the other angle.

4. The second angle in a triangle is three times as large as the first angle. The third angle is twice as large as the second angle. Find the measures of the angles.

5. Triangle *ABC* is an obtuse triangle. If m∠A = 23° and ∠B is the obtuse angle, find the largest possible measure for ∠C, to the nearest whole number.

(Answers are on page 201.)

Exterior Angles in Triangles

In the previous section, we looked at the **interior angles** of a triangle. These are the angles inside the triangle. But a triangle also has exterior angles.

The **exterior angles** are all the angles between one side of the triangle and the line you get by extending an adjacent side *outside* of the triangle. Look at the diagram below.

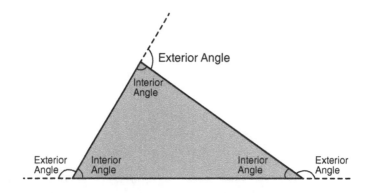

There is a relationship between the interior and exterior angles of a triangle. Look at triangle *ABC*.

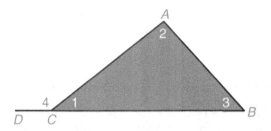

We already know that m∠1 + m∠2 + m∠3 = 180°.

Now we can also find the measure of an exterior angle: **The measure of an exterior angle of a triangle is equal to the sum of the measures of the two nonadjacent interior angles.**

1+2=3 MATH TALK!

The measure of an exterior angle (an outside angle) is equal to the measure of the two inside angles **not next to** the exterior angle (two nonadjacent interior angles).

According to our diagram: **m ∠ 4 = m ∠ 2 + m ∠ 3**.

Example:

> If m∠2 = 50° and m∠3 = 58°, find the measure of angle 4.

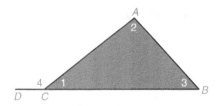

> Solution:
>
> m∠2 + m∠3 = m∠4
>
> 50 + 58 = m∠4
>
> 108° = m∠4

If the measure of angle 4 is the sum of angle 2 and angle 3, then it also makes sense that the measure of angle 4 is larger than both angle 2 and angle 3.

Look at our example again. We can see that m∠4 = 108°, which is larger than m∠2 = 50° and m∠3 = 58°. Another rule/theorem that exists then is

> **The measure of an exterior angle of a triangle is greater than either of its two nonadjacent interior angles.**

Examples:

1. In triangle PQR, m∠Q = 43° and m∠R = 71°. Find the measure of an exterior angle at P.

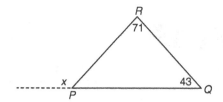

Solution:
$71 + 43 = x$
$114 = x$

The exterior angle is 114°.

2. Find the missing angles x and y.

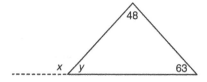

Solution:
There are a few ways to find these angles.

To find y first: $y + 48 + 63 = 180$
$$y + 111 = 180$$
$$y = 69°$$

and then $x + y = 180$ (since it's a linear pair)
so $180 - 69 = 111°$ for x.

Or find the exterior angle first:
$48 + 63 = x$ (exterior angle rule)
$111° = x$

To find y, use the fact that $x + y = 180$. So $180 - 111 = 69°$.

3. In triangle EFG, an exterior angle at G is represented by $8x + 15$. If the two nonadjacent angles are represented by $3x + 20$ and $4x + 5$, find the measures of the angles.

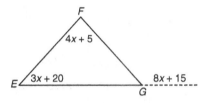

Solution:
$$3x + 20 + 4x + 5 = 8x + 15$$
$$7x + 25 = 8x + 15$$
$$\underline{-7x \qquad\quad -7x}$$
$$25 = x + 15$$
$$\underline{-15 \qquad -15}$$
$$10 = x$$

$\angle E = 3x + 20 = 3(10) + 20 = 50°$
$\angle F = 4x + 5 = 4(10) + 5 = 45°$
$\angle G = 8x + 15 = 8(10) + 15 = 95°$

The angles are 50°, 45°, and 95°.

4. Find the measure of an exterior angle at the base of an isosceles triangle whose vertex angle is 45°.

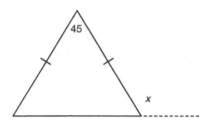

Solution:
The angles in a triangle add up to 180°.
So $180 - 45 = 135°$ left for two equal inside angles.
$135 \div 2 = 67.5°$ for each interior base angle.

Using the exterior angle rule:
$45 + 67.5 = x$
$x = 112.5°$

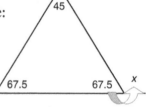

Another way to solve:
Notice interior angle + exterior angle = 180
$67.5 + x = 180$
$x = 112.5°$

BRAIN TICKLERS   Set # 37

1. In triangle *DEF*, m∠*D* = 40° and m∠*E* = 68°. Find the measure of the exterior angle at *F*.

2. Find the missing angles.

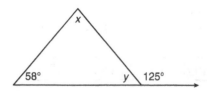

3. In triangle *LMN*, an exterior angle at *N* is represented by $10x - 18$. If ∠*L* and ∠*M* are represented by $6x + 4$ and $2x + 2$, find the measures of the angles inside the triangle.

4. In triangle ABC, $m\angle A = 2x$, $m\angle B = 4x$, and $m\angle C = 4x + 10$. Find the measure of the exterior angle at A.

5. If you do not know any of the angle measures in a triangle, can you find the sum of the exterior angles of the triangle? Explain.

(Answers are on page 201.)

Similar Figures

We know that **similar figures** have the same shape, but they are a different size. Two triangles are similar if the lengths of the corresponding sides are proportional and the corresponding angles are congruent (equal in measure).

Words	**Symbols**
Triangle *ABC* Is Similar to Triangle *DEF*	$\triangle ABC \sim \triangle DEF$

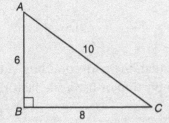

Since the two triangles are similar:

- **Corresponding angles are congruent**

  Measure of $\angle A$ is congruent/equal to the measure of $\angle D$

  $$m\angle A \cong m\angle D$$

  Measure of $\angle B$ is congruent/equal to the measure of $\angle E$

  $$m\angle B \cong m\angle E$$

  Measure of $\angle C$ is congruent/equal to the measure of $\angle F$

  $$m\angle C \cong m\angle F$$

  and

- **Corresponding sides have lengths that are proportional**

  $$\frac{\text{length of } AB}{\text{length of } DE} = \frac{\text{length of } BC}{\text{length of } EF} = \frac{\text{length of } AC}{\text{length of } DF}$$

  $$\frac{AB}{DE} = \frac{BC}{EF} = \frac{AC}{DF} \qquad \frac{6}{3} = \frac{8}{4} = \frac{10}{5}$$

Use what you know about proportions to help you solve problems involving similar figures.

Examples:

1. An isosceles triangle has a base of 15 centimeters and legs measuring 25 centimeters. How long are the legs of a similar triangle with a base measuring 37.5 centimeters?

Solution:

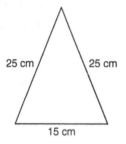 

Since the triangles are similar, the corresponding sides are proportional. Set up a proportion or use scale factor.

$$\frac{\text{leg}}{\text{base}} \qquad \frac{25}{15} = \frac{x}{37.5}$$

$$15x = 25(37.5)$$

$$\frac{15x}{15} = \frac{937.5}{15}$$

$$x = 62.5$$

$$\overset{\times\,2.5}{\frac{25}{15}} = \frac{x}{37.5}$$
$$\underset{\times\,2.5}{}$$

$$x = 62.5$$

The legs of the similar triangle are 62.5 centimeters.

2. A rectangle is 15 centimeters long and 8 centimeters wide. A similar rectangle is 4.5 centimeters wide and x centimeters long. How long is the similar figure?

Solution:

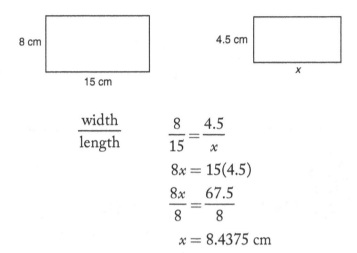

$$\frac{\text{width}}{\text{length}} \qquad \frac{8}{15} = \frac{4.5}{x}$$

$$8x = 15(4.5)$$

$$\frac{8x}{8} = \frac{67.5}{8}$$

$$x = 8.4375 \text{ cm}$$

The length of the similar rectangle is 8.4375 centimeters.

3. Since the water level at the lake has risen, a new boat ramp needs to be made. A ramp is built by putting a triangle on top of the trapezoid base. How long is the new ramp?

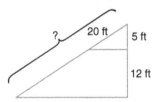

Solution:
Be careful when setting up your proportion! Remember, you need "like" sides.

$$\frac{\text{side of triangle}}{\text{hypotenuse of triangle}} \qquad \frac{5}{20} = \frac{17}{x}$$

$$5x = 20(17)$$

$$\frac{5x}{5} = \frac{340}{5}$$

$$x = 68$$

The length of the new ramp is 68 feet.

Caution–Major Mistake Territory!

When comparing similar triangles, draw the triangles **separately** to see the corresponding sides.

Be VERY careful setting up "like" sides.

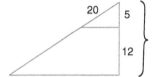

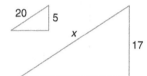

BRAIN TICKLERS   Set # 38

1. Which triangles are similar? Justify your reasoning.

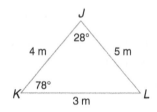

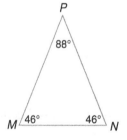

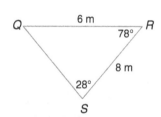

2. A building is 20 feet tall, and it casts a shadow that is 35 feet long. At the same time, a nearby tree is 9 feet tall. If the figures are similar, how long is the tree's shadow? Round your answer to the nearest tenth of a foot.

3. Chris wants to know the distance across his pond. He drew the following picture and labeled it with his measurements. How far is it across his pond?

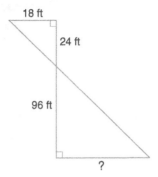

4. The model of a 27-foot-tall house was made using the scale 2 inches : 3 feet. Is the height of the model of the house 9 inches? Explain why or why not, and justify your reasoning.

(Answers are on page 202.)

Two Parallel Lines Cut by a Transversal

Parallel lines are lines in a plane that never cross or meet. Line 1 and Line 2 are parallel lines.

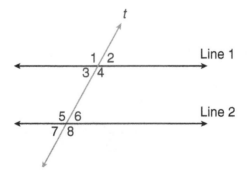

A **transversal** (represented by line *t* in the diagram above) is a line that intersects or crosses two or more lines in the same plane. When two parallel lines are cut by a transversal, special angle relationships occur.

These are congruent (equal in measure) angles that are formed when two parallel lines are cut by a transversal.

Vertical angles   **Corresponding angles**
$\angle 1 \cong \angle 4$ $\angle 1 \cong \angle 5$
$\angle 2 \cong \angle 3$ $\angle 2 \cong \angle 6$
$\angle 5 \cong \angle 8$ $\angle 3 \cong \angle 7$
$\angle 6 \cong \angle 7$ $\angle 4 \cong \angle 8$

Alternate interior angles   **Alternate exterior angles**
$\angle 3 \cong \angle 6$ $\angle 1 \cong \angle 8$
$\angle 4 \cong \angle 5$ $\angle 2 \cong \angle 7$

There are also a lot of **supplementary angles** that show up when two parallel lines are cut by a transversal. Remember that supplementary angles are two angles that add up to 180°.

These are some of the supplementary angles:

$\angle 1 + \angle 2 = 180$ $\angle 3 + \angle 4 = 180$ $\angle 5 + \angle 6 = 180$
$\angle 7 + \angle 8 = 180$ $\angle 1 + \angle 3 = 180$ $\angle 2 + \angle 4 = 180$
$\angle 5 + \angle 7 = 180$ $\angle 6 + \angle 8 = 180$

Knowing these angle relationships will help you find any angle measure if you are given one angle to start.

Examples:

1. If m$\angle 1 = 110°$, find the remaining angles.

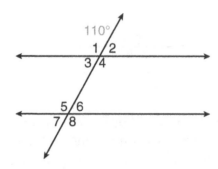

Solution:

∠4 = 110° because ∠1 and ∠4 are vertical angles.

∠5 = 110° because ∠4 and ∠5 are alternate interior (or ∠1 and ∠5 are corresponding) angles.

∠8 = 110° because ∠5 and ∠8 are vertical (or ∠4 and ∠8 are corresponding) angles.

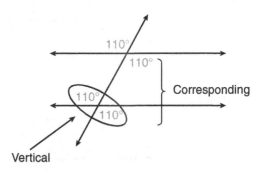

Corresponding angles are in the same position according to the transversal.

Vertical angles are opposite each other when two lines cross.

If m∠1 = 110°, ∠1 + ∠2 = 180°. So m∠2 = 70°. If m∠2 = 70°, then

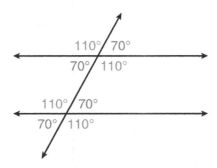

∠3 = 70° because ∠2 and ∠3 are vertical angles.

∠6 = 70° because ∠3 and ∠6 are alternate interior (or ∠2 and ∠6 are corresponding) angles.

∠7 = 70° because ∠6 and ∠7 are vertical (or ∠3 and ∠7 are corresponding) angles.

Find ALL the missing angles. Line 1 is parallel to Line 2.

2.

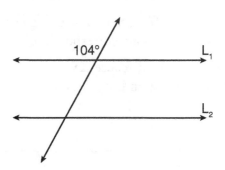

Solution:

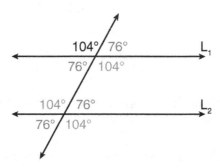

3.

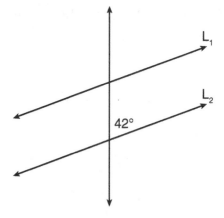

Solution:

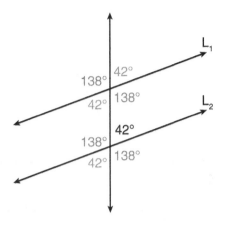

4.

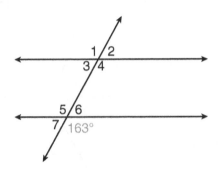

Name an angle vertical to 163°.

∠5 (opposite the angle)

Name an angle corresponding to 163°.

∠4 (same position along the transversal)

∠5 and ∠4 are congruent because they are what type of angles?

Alternate interior angles

What angle and 163° are alternate exterior angles?

∠1

Are angles 3 and 163° supplementary? Justify your answer.

Yes, they add up to 180°. Fill in the other angles until you get to ∠3.

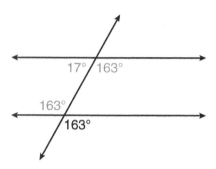

BRAIN BUSTER!

Use what you know about all angles and parallel lines to find all the missing angles.

5.

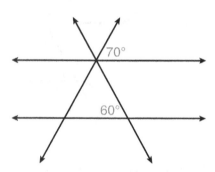

Solution:

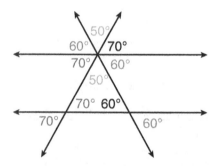

Start with 70°, and find vertical and alternate interior angles.

Start with 60°, and find vertical and alternate interior angles.

Look at the top three angles. 60 + 70 + missing angle add up to 180, so the missing angle is 50°.

BRAIN TICKLERS   Set # 39

1. In the following diagram, Line *a* is parallel to Line *b*.

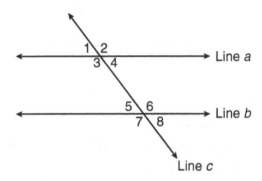

Line *c*

a. Name two sets of alternate interior angles.

b. Name two sets of alternate exterior angles.

c. List all the pairs of corresponding angles.

d. Name all the angles supplementary to ∠5.

e. What is true about the measures of ∠2 and ∠3? What type of angles are they?

f. Explain why m∠1 ≅ m∠4, m∠1 ≅ m∠5, and m∠1 ≅ m∠8.

g. If m∠1 = 30°, find all the missing angles.

h. Line *a* and Line *b* are what type of lines?

i. What is the name of Line *c*?

j. Is it ever possible for all the angles to equal 90°? Draw a picture to support your answer.

(Answers are on page 202.)

Parallel Lines Cut by a Transversal– Algebraically

In the previous section, you learned angle relationships when two parallel lines are cut by a transversal. Let's apply that knowledge to find the angle measures when they are given to us algebraically!

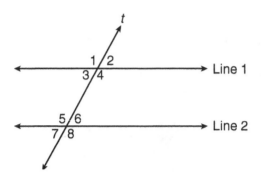

Important concepts to keep in mind when setting up your equations:

- Vertical angles are congruent.
- Corresponding angles are congruent.
- Alternate interior angles are congruent.
- Alternate exterior angles are congruent.
- Angles that are supplementary add up to 180°.

To solve for an angle algebraically:

1. Write an equation based on what you know about the angle relationship.

2. Solve for x.

3. Find the angle measure that is asked.

Examples:

For the following, assume both lines are parallel.

1. Find the angles.

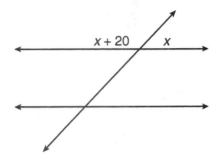

Solution:
Since the angles are supplementary

$$x + 20 + x = 180$$

Now solve:

$$x + 20 + x = 180$$
$$2x + 20 = 180$$
$$2x = 160$$
$$x = 80$$

So the angles are 80° and 100°.

2. Solve for x.

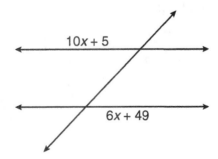

Solution:

Since the angles are alternate exteriors, they are equal to each other. The equation would be $10x + 5 = 6x + 49$.

$$10x + 5 = 6x + 49$$
$$\underline{-6x \qquad -6x}$$
$$4x + 5 = 49$$
$$4x = 44$$
$$x = 11 \qquad \boxed{\text{The solution is } x = 11.}$$

Does this check?

$$10x + 5 = 10(11) + 5 = 115°$$
$$6x + 49 = 6(11) + 49 = 115°$$

Yes! They are congruent.

3. Solve for x.

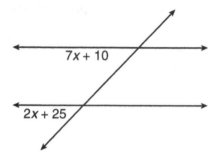

Solution:

Since the angles are corresponding angles, they are equal to each other. The equation would be $7x + 10 = 2x + 25$.

$$7x + 10 = 2x + 25$$
$$\underline{-2x \qquad -2x}$$
$$5x + 10 = 25$$
$$5x = 15$$
$$x = 3 \qquad \boxed{\text{The solution is } x = 3.}$$

BRAIN TICKLERS   Set # 40

For all examples, assume the lines are parallel.

1. In the diagram below, L_1 is parallel to L_2.

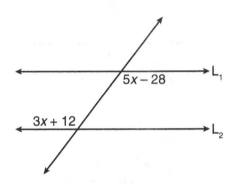

5x − 28

3x + 12

L_1

L_2

   a. What is the relationship between the two angles?

   b. Solve for x.

   c. What is the measure of the angle $3x + 12$?

   d. What is the measure of the supplement of $3x + 12$?

2. Solve for x. Explain your reasoning.

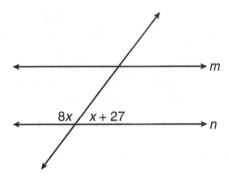

8x   x + 27

m

n

3. Find the measures of the angles. What type of angles are these?

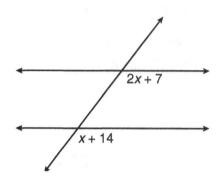

2x + 7

x + 14

4.

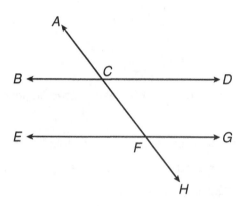

The measure of ∠ACD = 4x + 10, and the measure of ∠BCF = x + 115. Find the measures of the angles. What is the measure of ∠DCF? The measure of ∠EFH?

5. Transversal $\overleftrightarrow{EH}$ intersects parallel lines $\overleftrightarrow{AB}$ and $\overleftrightarrow{CD}$ at F and G, respectively. If m∠DGH = x° and the measure of ∠BFE = 2x − 30°, find the value of x.

(Answers are on page 203.)

BRAIN TICKLERS—THE ANSWERS
Set # 34, pages 174-175
1. 46°
2. 117°
3. $90 - 5x$
4. $180 - 7x$
5. 46°
6. 65°; supplementary angles
7. a. $\angle CPB$ (or $\angle BPC$); vertical angles
   b. $\angle APC$. $\angle APC$ and $\angle APD$ make a straight line.

Set # 35, page 177
1. $x = 13; 45°, 135°$
2. $x = 5; 25°, 65°$
3. $x = 20; 20°, 160°$
4. 58° and 122°
5. 60°

Set # 36, page 180
1. 66.5°
2. 80°, 60°, 40°
3. 38°
4. 18°, 54°, 108°
5. 66°

Set # 37, pages 184-185
1. 108°
2. $x = 67°, y = 55°$
3. $\angle L = 76°, \angle M = 26°, \angle N = 78°$
4. 146°
5. Yes, the sum of the exterior angles = 360° because a circle can be made around the triangle.

Set # 38, pages 188–189

1. Triangles *JKL* and *SRQ*. The angles are the same, and the sides are proportional.

2. 15.8 feet

3. 72 feet

4. No, the height should be 18 inches to keep the 2:3 ratio.

Set # 39, page 195

1. a. $\angle 4$ and $\angle 5$, $\angle 3$ and $\angle 6$

   b. $\angle 1$ and $\angle 8$, $\angle 2$ and $\angle 7$

   c. $\angle 1$ and $\angle 5$, $\angle 2$ and $\angle 6$, $\angle 3$ and $\angle 7$, $\angle 4$ and $\angle 8$

   d. $\angle 2$, $\angle 3$, $\angle 6$, $\angle 7$

   e. $\angle 2$ and $\angle 3$ are congruent. They are vertical angles.

   f. Vertical angles, corresponding angles, alternate exterior angles.

   g. $\angle 1 = 30°$, $\angle 2 = 150°$, $\angle 3 = 150°$, $\angle 4 = 30°$, $\angle 5 = 30°$, $\angle 6 = 150°$, $\angle 7 = 150°$, $\angle 8 = 30°$

   h. Parallel lines

   i. Transversal

   j. Yes, they have to be perpendicular.

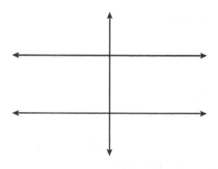

Set # 40, pages 199–200

1. a. They are alternate interior angles.

   b. $x = 20$

   c. $72°$

   d. $108°$

2. The angles are supplementary and add up to 180. $x = 17$.

3. Corresponding angles, each 21°.

4. $\angle ACD = 150°$, $\angle BCF = 150°$, $\angle DCF = 30°$, $\angle EFH = 150°$

5. $x = 70°$

Chapter 7

Transformations

Translations

A translation is the same as a slide. If you are sitting in a movie theater and decide to move three seats left and four seats forward, you just did a translation! You are still looking at the movie screen; you are just sitting in a different seat. How does this work on the coordinate plane? It's painless, you will see!

Examples:

Translate each of the following points.

$A(3, 5)$ 4 units left, 3 units up
The new point is $A'(-1, 8)$.

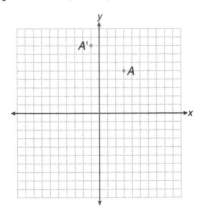

$B(-4, 6)$ 6 units right, 2 units down
The new point is $B'(2, 4)$.

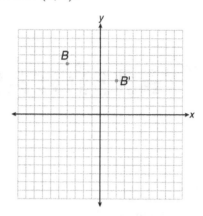

$C(-3, -7)$ 8 units right, 3 units up
The new point is $C'(5, -4)$.

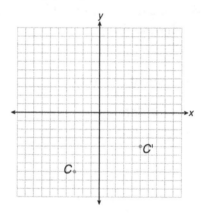

$D(4, -3)$ 7 units left, 3 units down
The new point is $D'(-3, -6)$.

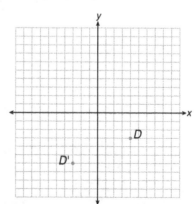

1+2=3 MATH TALK!

To translate, just move your original point . . .

SLIDE!

Left or right (changes the *x*-value)

Up or down (changes the *y*-value)

Translate each of the following:

Given triangle $A(-6, -3)$ $B(-3, -3)$ $C(-6, -7)$.

Translate ABC 4 units right and 6 units up. Label the new triangle $A'B'C'$ and the coordinates.

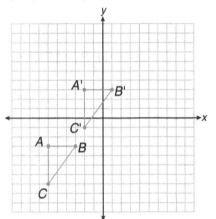

Solution:
$A(-6, -3)$ becomes $A'(-2, 3)$.
$B(-3, -3)$ becomes $B'(1, 3)$.
$C(-6, -7)$ becomes $C'(-2, -1)$.

Given triangle $A(2, 1)$ $B(5, 6)$ $C(6, 3)$ and triangle $A'(-6, -1)$ $B'(-3, 4)$ $C'(-2, 1)$.

What translation would move triangle ABC to triangle $A'B'C'$?

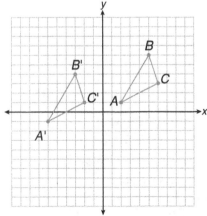

Solution:
Translating triangle ABC 8 units left and 2 units down.

Notice when you do a translation, the original shape and the new shape are the same shape, same size, and still have the same angle measures. These are **congruent** triangles.

MATH TALK!

You can write a rule for each translation you do. Remember: Moving right is positive and left is negative for x. Moving up is positive and down is negative for y.

Examples:

Move right 6, up 5	$(x, y) \rightarrow (x + 6, y + 5)$
Move right 7, down 4	$(x, y) \rightarrow (x + 7, y - 4)$
Move left 2, up 6	$(x, y) \rightarrow (x - 2, y + 6)$
Move left 4, down 8	$(x, y) \rightarrow (x - 4, y - 8)$

BRAIN TICKLERS   Set # 41

1. a. Graph and label triangle BCD, where $B(-5, 2)$, $C(-2, 4)$, and $D(-1, 1)$.

   b. Graph and state the coordinates of $B'C'D'$ after translating BCD $(x + 3, y - 4)$.

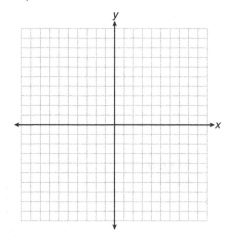

2. What translation maps $A(1, 5)$, $B(2, 8)$, $C(3, 5)$ onto $A'(6, 1)$, $B'(7, 4)$, and $C'(8, 1)$?

3. When translated, $A(-5, -12)$ has coordinates $A'(6, 10)$. Describe the translation.

4. a. Graph quadrilateral $W(0, 0)$, $X(2, 2)$, $Y(2, -1)$, and $Z(0, -3)$.

b. Translate $WXYZ$ 5 units left and 4 units up. Label the new quadrilateral $W'X'Y'Z'$.

c. Translate $W'X'Y'Z'$ 8 units right. Label this $W''X''Y''Z''$.

d. Are all three quadrilaterals similar or congruent? Explain.

e. Name a translation that would move $WXYZ$ to $W''X''Y''Z''$.

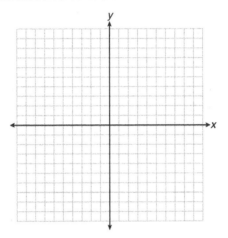

(Answers are on page 230.)

Reflections

A **reflection** is the same as a mirror image or a flip. In real life, you will see your reflection in a mirror or in a lake, but in math class, we usually do reflections on graph paper. Hence, we will look at how to reflect over a line.

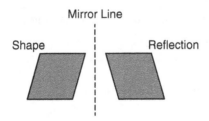

An object and its reflection will have the **same shape and same size**, but the **figures will face in opposite directions**. The shapes are **congruent**.

Reflection over the x-axis
Reflecting a point over the x-axis

To reflect over the x-axis, count how many lines you are away from the axis (above or below), and then count the same distance away from the x-axis.

Example:

$A(3, 5)$ becomes $A'(3, -5)$. A is 5 units above the x-axis, so count 5 units below the x-axis.

$B(-6, 2)$ becomes $B'(-6, -2)$.

$C(-4, -6)$ becomes $C'(-4, 6)$. C is 6 units below the x-axis, so count 6 units above the x-axis.

$D(-2, 8)$ becomes $D'(-2, -8)$.

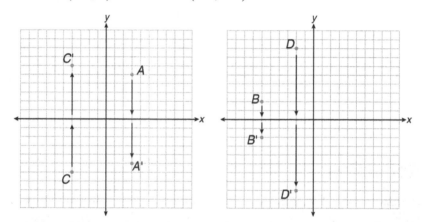

1+2=3 MATH TALK!

When you reflect over the x-axis, you negate the y-value.

(x, y) becomes $(x, -y)$.

Reflecting a shape over the x-axis
Examples:

1. Given triangle $A(-6, -3)$ $B(-3, -3)$ $C(-6, -7)$.

   a. Reflect triangle ABC over the x-axis and label the new triangle $A'B'C'$.

   b. Are the two triangles similar or congruent? Explain.

Solution:

a. Move one point at a time, count the distance from the x-axis, and then count the same distance away from the x-axis.

$A(-6, -3)$ becomes $A'(-6, 3)$.
$B(-3, -3)$ becomes $B'(-3, 3)$.
$C(-6, -7)$ becomes $C'(-6, 7)$. Be sure to label all of your points!

b. The shapes are congruent. The triangles are the same shape and the same size.

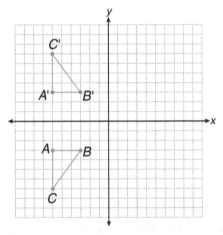

2. Given triangle $A(2, 1)$ $B(5, 6)$ $C(6, 3)$.

Reflect triangle ABC over the x-axis.
Label the new triangle $A'B'C'$, and state the coordinates.

Solution:
$A(2, 1)$ becomes $A'(2, -1)$.
$B(5, 6)$ becomes $B'(5, -6)$.
$C(6, 3)$ becomes $C'(6, -3)$.

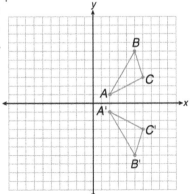

3. Given triangle $J(2, -2)$ $K(5, 6)$ $L(9, -1)$.

Reflect triangle JKL over the x-axis.
Label the new triangle $J'K'L'$.

Solution:
It is possible for some shapes to have points above and below the x-axis. Pay special attention to the direction you need to move each point.

$J'(2, 2)$ $K'(5, -6)$ $L'(9, 1)$

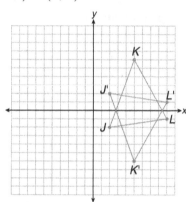

Reflecting over the y-axis
Reflecting a point over the y-axis

Plot each point and reflect each point over the y-axis. Use the same strategy as you did above, just make sure you are moving over the y-axis!

$A(3, 5)$ becomes $A'(-3, 5)$. $B(-4, 6)$ becomes $B'(4, 6)$.
$C(-3, -7)$ becomes $C'(3, -7)$. $D(4, -3)$ becomes $D'(-4, -3)$.

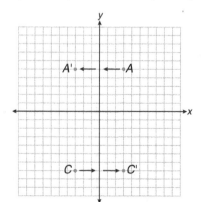
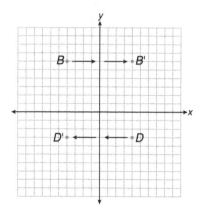

1+2=3 **MATH TALK!**

When you reflect over the *y*-axis, you negate the *x*-value.

(x, y) becomes $(-x, y)$.

Reflecting a shape over the y-axis
Examples:

Given triangle $A(-6, -3)$ $B(-3, -3)$ $C(-6, -7)$.

Reflect triangle ABC over the *y*-axis.
Label the new triangle $A'B'C'$, and state the coordinates.

Solution:
$A(-6, -3)$ becomes
$A'(6, -3)$.
$B(-3, -3)$ becomes
$B'(3, -3)$.
$C(-6, -7)$ becomes
$C'(6, -7)$.

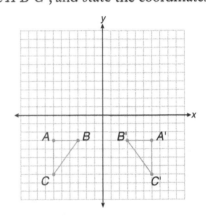

Given triangle $A(2, 1)$ $B(5, 6)$ $C(6, 3)$.

Reflect triangle ABC over the *y*-axis.
Label the new triangle $A'B'C'$.

Solution:
$A(2, 1)$ becomes
$A'(-2, 1)$.
$B(5, 6)$ becomes
$B'(-5, 6)$.
$C(6, 3)$ becomes
$C'(-6, 3)$.

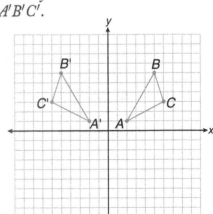

BRAIN BUSTER!

Use what you have learned about reflections to do this problem!

Graph trapezoid *BIRD* with vertices $B(1, 1)$, $I(2, 4)$, $R(6, 4)$, and $D(7, 1)$.

a. Find the coordinates $B'I'R'D'$ after reflecting *BIRD* over the *x*-axis.

b. Find the coordinates of $B''I''R''D''$ after reflecting $B'I'R'D'$ over the *y*-axis.

Solutions:

a. $B'(1, -1), I'(2, -4), R'(6, -4), D'(7, -1)$

b. Make sure you used your reflected shape to now reflect over the *y*-axis!

$$B''(-1, -1), I''(-2, -4), R''(-6, -4), D''(-7, -1)$$

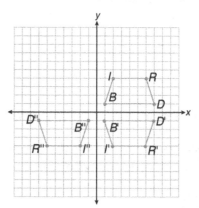

Look at all of the shapes that you reflected, whether over the *x*-axis or the *y*-axis. What did you notice about their shapes and sizes? When reflecting shapes, the shape and the size stay the same as the original figure. That means that the shapes are **congruent**.

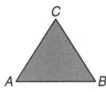 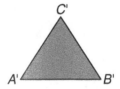

1+2=3 MATH TALK!

Two shapes are **congruent** when they have exactly the same shape and size.

Words: Triangle *ABC* is congruent to triangle *A'B'C'*.

Symbols: $\triangle ABC \cong \triangle A'B'C'$.

The sides and angles that match each other are called **corresponding sides** and **corresponding angles.**

Angle *A* corresponds to angle *A'*.
Angle *B* corresponds to angle *B'*.
Angle *C* corresponds to angle *C'*.
Side *AC* corresponds to side *A'C'*.
Side *AB* corresponds to side *A'B'*.
Side *CB* corresponds to side *C'B'*.

BRAIN TICKLERS   Set # 42

1. Graph triangle $A(-5, -1)\ B(-7, -4)\ C(-2, -6)$.

   Reflect triangle *ABC* over the *x*-axis. Label the new triangle *A'B'C'*, and state the coordinates.

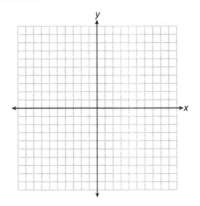

2. Graph triangle $J(-6, 3)$ $K(-4, 7)$ $L(2, 6)$.

Reflect triangle *JKL* over the *y*-axis. Label and state the coordinates of triangle *J'K'L'*. Are these triangles similar or congruent? Explain.

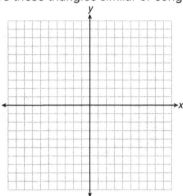

3. Graph parallelogram *JUNE* with vertices $J(2, -2)$, $U(6, -2)$, $N(8, -5)$, and $E(4, -5)$.

Graph and state the coordinates of *J'U'N'E'* after a reflection of *JUNE* over the *x*-axis.

Graph and state the coordinates of *J"U"N"E"*, the image of *J'U'N'E'* after a reflection over the *y*-axis.

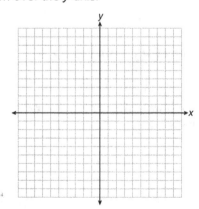

4. Graph triangle *USA* with vertices *U*(0, 4), *S*(4, 4), *A*(4, 0).

    a. Reflect triangle *USA* over the *y*-axis. Label *U'S'A'*.

    b. Reflect triangle *USA* over the *x*-axis. Label *U"S"A"*.

    c. Reflect *U'S'A'* over the *x*-axis. Label *U'''S'''A'''*.

    d. What is the area of the square formed by the four triangles?

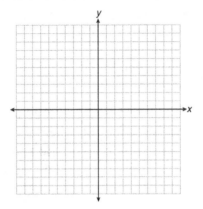

(Answers are on pages 231–232.)

Rotations

What do a Ferris wheel, a merry-go-round, and a tire all have in common? They rotate or turn! A **rotation** turns a figure around a fixed point. The original object and its rotation will have the same shape and size, but they may be turned in a different direction.

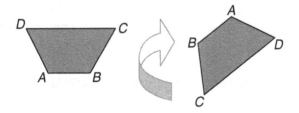

The diagram above shows *ABCD* rotated in a **clockwise** direction.

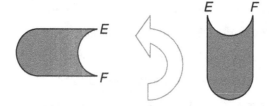

The diagram here shows the shape rotated in a **counterclockwise** direction.

Notice in both examples, the pictures stay the same shape and size (congruent) but they are turned.

How do you rotate an object? If someone said to you "rotate" or "turn," some people would turn right, whereas others would turn left. Therefore, a rule was created to help people know that a rotation in math would always go in a certain direction, unless otherwise stated.

BIG IDEA!

Unless otherwise stated, a rotation always turns an object **COUNTERCLOCKWISE around the origin (0, 0).**

Let's look at the rules for a rotation.

Rules for rotations (counterclockwise!)

Rotation	Rule	Words
90°	$(x, y) \rightarrow (-y, x)$	(negate y, keep x)
180°	$(x, y) \rightarrow (-x, -y)$	(negate x, negate y)
270°	$(x, y) \rightarrow (y, -x)$	(keep y, negate x)

NOTICE: These rules always start from the ORIGINAL (x, y) coordinate.

Examples:

1. Rotate the following points 90°.

   x y
   $A(2, 4) \rightarrow$ need $(-y, x) \rightarrow A'(-4, 2)$
   In symbols: $R_{90°}(2, 4) = (-4, 2)$

   $B(3, -5)$ becomes $B'(5, 3)$
   In symbols: $R_{90°}(3, -5) = (5, 3)$

   $C(-4, -5)$ becomes $C'(5, -4)$
   $D(-4, 7)$ becomes $D'(-7, -4)$

1+2=3 MATH TALK!

For a rotation of 90°, take the opposite of the y-value and write it first, keep the original x-value and write it second.

2. Rotate the following 180°.

   x y
   $A(2, 4) \rightarrow$ need $(-x, -y) \rightarrow A'(-2, -4)$
   In symbols: $R_{180°}(2, 4) = (-2, -4)$

   $B(3, -5)$ becomes $B'(-3, 5)$
   In symbols: $R_{180°}(3, -5) = (-3, 5)$

   $C(-4, -5)$ becomes $C'(4, 5)$
   $D(-4, 7)$ becomes $D'(4, -7)$

1+2=3 MATH TALK!

For a rotation of 180°, keep (x, y) in the same spot, just take the opposite of the x-value and the opposite of the y-value.

3. Rotate the following 270°.

$$x \ y$$
$A(2,4) \rightarrow$ need $(y,-x) \rightarrow A'(4,-2)$
In symbols: $R_{270°}(2,4) = (4,-2)$

$B(3,-5)$ becomes $B'(-5,-3)$
In symbols: $R_{270°}(3,-5) = (-5,-3)$

$C(-4,-5)$ becomes $C'(-5,4)$

$D(-4,7)$ becomes $D'(7,4)$

1+2=3 MATH TALK!

For a rotation of 270°, keep the original *y*-value and write it first, take the opposite of the original *x*-value and write it second.

Why do we not rotate 360°? Turning 360° will give you the original shape! No need! If you wanted to rotate from 270° to 360°, use the rule for 90° on the 270° coordinate $(x,y) \rightarrow (-y,x)$ (because $270 + 90 = 360$), and you will see that you will just get the original shape back.

Example:

$$\qquad\quad 90° \qquad\quad 180° \qquad\quad 270° \qquad 360°$$
$$A(1,4) \rightarrow (-4,1) \rightarrow (-1,-4) \rightarrow (4,-1) \rightarrow (1,4)$$

SHORTCUT!

If you want to remember only ONE rule, **remember the 90° rule, $(x, y) \rightarrow (-y, x)$**, and use this rule repeatedly until you get to the rotation you need!

Look at the example again!

$$\qquad\quad 90° \qquad\quad 180° \qquad\quad 270° \qquad 360°$$
$$A(1, 4) \rightarrow (-4, 1) \rightarrow (-1, -4) \rightarrow (4, -1) \rightarrow (1, 4)$$
$$\qquad (-y, x) \qquad (-y, x) \qquad (-y, x) \qquad (-y, x)$$

This makes sense because you are just adding 90° every time!

At this point you need to decide what works best for you. Some people like to memorize one rule and use it over and over, some people prefer the three rules. Just make sure you understand what works for you!

Let's rotate a shape!

Examples:

1. Given triangle $A(-6, 7)$ $B(-2, 3)$ $C(-7, 1)$.
   Graph and state the coordinates of triangle $A'B'C'$ after a $90°$ counterclockwise rotation.

   Solution:
   $A(-6, 7)$ becomes
   $A'(-7, -6)$.
   $B(-2, 3)$ becomes
   $B'(-3, -2)$.
   $C(-7, 1)$ becomes
   $C'(-1, -7)$.

   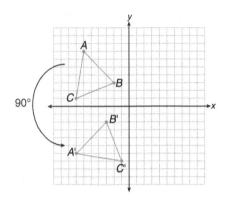

2. Given triangle $P(2, 3)$ $Q(5, 8)$ $R(8, 5)$.
   Graph and state the coordinates of $P'Q'R'$ after a rotation of $270°$ about the origin.

   Solution:
   $P(2, 3)$ becomes
   $P'(3, -2)$.
   $Q(5, 8)$ becomes
   $Q'(8, -5)$.
   $R(8, 5)$ becomes
   $R'(5, -8)$.

   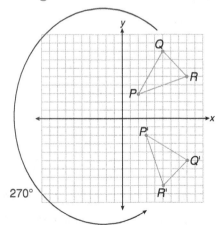

BRAIN TICKLERS   Set # 43

1. Complete each rule for finding the image of any point (x, y) under the given rotation.

   a. $R_{90°}$ about the origin:

   (x, y) → (,   )

   b. $R_{180°}$ about the origin:

   (x, y) → (,   )

   c. $R_{270°}$ about the origin:

   (x, y) → (,   )

   d. $R_{360°}$ about the origin:

   (x, y) → (,   )

2. Complete the table using your rules.

Starting point	90° rotation	180° rotation	270° rotation	360° rotation
A (3, 4)				
B (−5, 2)				
C (2, −8)				
D (−1, −6)				
E (0, 7)				

3. What is the image of $(-7, 3)$ under a 180° counterclockwise rotation about the origin?

4. What are the coordinates of $(6, 2)$ under a 90° clockwise rotation about the origin?

5. What is the image of $R_{90°}(1, 5)$?

6. **a.** Graph the triangle $A(3, -6)$ $B(5, -2)$ $C(7, -6)$ on the axes provided.

   b. Graph and state the coordinates of $A'B'C'$, the image of triangle ABC, after a 90° rotation counterclockwise.

   c. Graph and state the coordinates of $A''B''C''$, the image of triangle ABC, after a 180° rotation about the origin.

d. Graph and state the coordinates of triangle A'''B'''C''', the image of triangle ABC, after a rotation of 270° about the origin.

e. Are these triangles similar or congruent? Explain your reasoning.

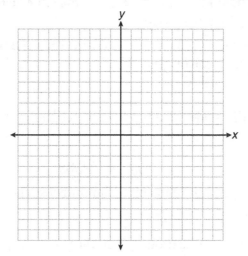

(Answers are on page 233.)

Dilations

A blueprint of a house compared to the actual house. Your pupil dilates when the eye doctor puts drops in your eyes. These are examples of dilations.

A **dilation** is a transformation that produces an image that is the same shape but a different size. An easy way to remember dilation is that it stretches or shrinks an object.

This is the first time a transformation CHANGES THE SIZE of the shape. These shapes will be similar. Shapes that are **similar** have the same shape, but are a different size.

How do you change the size of a shape?

> A **scale factor** is used to create an **enlargement**, an image larger than the original image. If the scale factor is greater than 1, the image will stretch.
>
> A scale factor is also used to create a **reduction**, a smaller image than the original. If the scale factor is between 0 and 1, the image will shrink.

A dilation with a scale factor of k, whose center of dilation is the origin, may be written like this:

$$D_k(x, y) = (kx, ky)$$

k is multiplied by each number, x and y.

Examples:

1. Perform a dilation of 2 on the point $A(3, 5)$.

   Steps: Multiply both coordinates by the scale factor. $(3 \times 2, 5 \times 2)$
   Simplify and label the new point. $A'(6, 10)$

2. Perform a dilation of ½ on the point $B(12, 6)$.

   Steps: Multiply both coordinates by ½. $(12 \times ½, 6\ 3\ ½)$
   Simplify and label the new point. $B'(6, 3)$

3. Point $C(15, 30)$ becomes $C'(3, 6)$ using what scale factor?

   Solution: Be careful here! **Scale factor must always be MULTIPLIED!** The scale factor is $^1/_5$.

More examples—Dilations on the coordinate plane:

1. Given triangle $A(1, 3)\ B(2, -1)\ C(-1, -1)$.

   a. Graph and state the coordinates of triangle $A'B'C'$ after a dilation with a scale factor of 3.

   b. Are the two triangles similar or congruent? Explain.

Solution:

a. $A(1, 3)$ becomes $A'(3, 9)$.

  $B(2, -1)$ becomes $B'(6, -3)$.

  $C(-1, -1)$ becomes $C'(-3, -3)$.

b. The two triangles are similar because they are the same shape but a different size.

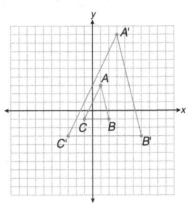

2. Given quadrilateral $A(0, 6)$ $B(4, -2)$ $C(0, -6)$ $D(-6, -2)$.

   Graph and state the coordinates of $A'B'C'D'$ after a dilation with a scale factor of ½.

   Are these shapes similar or congruent? Explain.

   Solution:
   $A'(0, 3)$ $B'(2, -1)$ $C'(0, -3)$ $D'(-3, -1)$

   The shapes are similar because they are the same shape but a different size.

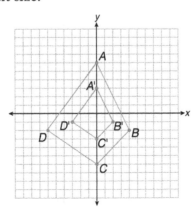

BRAIN TICKLERS   Set # 44

1. Triangle *LMN* has vertices *L*(8, 2), *M*(10, 8), *N*(4, 6).

   a. Graph triangle *LMN*.

   b. Graph and state the coordinates of *L'M'N'*, the image of *LMN* after a dilation with a scale factor of $\frac{1}{2}$.

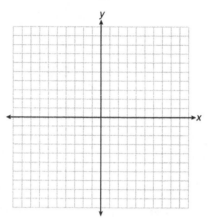

2. Triangle *PQR* has coordinates *P*(−2, 2), *Q*(3, 2), and *R*(0, −2).

   a. Graph triangle *PQR*.

   b. Graph and state the coordinates *P'Q'R'*, the image of *PQR* after a dilation of 3.

   c. Graph and state the coordinates of the image of triangle *PQR* after a dilation with a scale factor of $\frac{1}{2}$. Label this *P"Q"R"*.

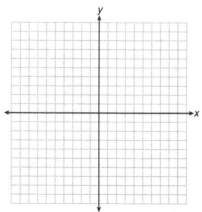

(Answers are on page 234.)

Series of Transformations

There are four main transformations: translations, reflections, rotations, and dilations. Let's quickly review what each one means.

Type	Nickname	Characteristics	Example
Translation	Slide	Same shape Same size Object moved to a different position, facing same direction Shapes congruent	B B
Reflection	Flip Mirror image	Same shape Same size Object flipped over Shapes congruent	B⦙ᗺ
Rotation	Turn	Same shape Same size Object turned Shapes congruent	B ℬ
Dilation	Shrink Enlarge	Same Shape Different size Object faces same direction Shapes similar	B B

Instead of using just one transformation on an object, let's use two or three. What is most important when doing this? READ! Make sure you are doing the correct transformation. Refer to the chart as needed.

Examples:

1. Given triangle CAT, in which $C(-1, -2)$ $A(-2, 4)$ and $T(-5, -3)$. Reflect over the y-axis and then rotate $90°$ counterclockwise. State all coordinates and label the images $C'A'T'$ and $C''A''T''$.

   Solution:

   $r_{y\text{-axis}}$
   $C'(1, -2), A'(2, 4), T'(5, -3)$

   $R_{90°}$
   $C''(2, 1), A''(-4, 2), T''(3, 5)$

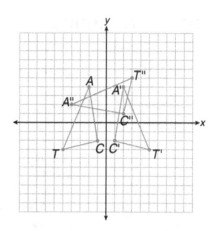

2. Graph triangle A(6, −2) B(10, 5) C(2, 8). Rotate 270°, and then dilate the new triangle by a scale factor of ½. Label the triangles $A'B'C'$ and $A''B''C''$. Which triangles are similar and which triangles are congruent? Justify your reasoning.

Solution:

$R_{270°}$

$A'(-2, -6), B'(5, -10), C'(8, -2)$

$A''(-1, -3), B''(2.5, -5), C''(4, -1)$

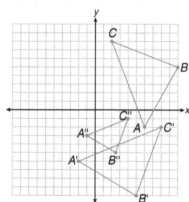

Triangles ABC and $A'B'C'$ are congruent because they are the same shape and same size. Triangles ABC and $A'B'C'$ are both similar to triangle $A''B''C''$, because they are the same shape but double the size of $A''B''C''$.

BRAIN TICKLERS   Set # 45

Do each of the following on separate graph paper.

1. Graph quadrilateral $A(2, 1)$ $B(3, 7)$ $C(-2, 8)$ $D(-1, 0)$. Rotate *ABCD* 180° counterclockwise about the origin and label this $A'B'C'D'$. Then translate $A'B'C'D'$ 5 units to the right and 4 units up. Label this quadrilateral $A''B''C''D''$. State all your coordinates.

Are these figures similar or congruent? Justify your answer.

2. Graph triangle $X(2, 8)$ $Y(5, 10)$ $Z(5, 2)$. Reflect *XYZ* over the *x*-axis and label this $X'Y'Z'$. Then reflect $X'Y'Z'$ over the *y*-axis and label it $X''Y''Z''$.

Explain what transformation you could then use to move Z'' to $Z'''(-1, 5)$.

3. Explain what two series of transformations has occurred to move triangle *ABC* to triangle $A''B''C''$.

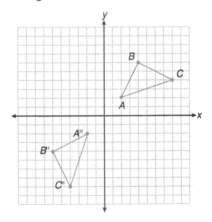

(Answers are on page 235.)

BRAIN TICKLERS—THE ANSWERS
Set # 41, pages 208-209

1. a. See graph.

   b. $B'(-2, -2)$ $C'(1, 0)$ $D'(2, -3)$

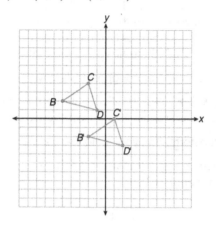

2. 5 right, 4 down or $(x + 5, y - 4)$

3. 11 right, 22 up $(x + 11, y + 22)$

4. a. See graph.

   b. $W'(-5, 4), X'(-3, 6), Y'(-3, 3), Z'(-5, 1)$

   c. $W''(3, 4), X''(5, 6), Y''(5, 3), Z''(3, 1)$

   d. Congruent, same shape and size.

   e. 3 right, 4 up

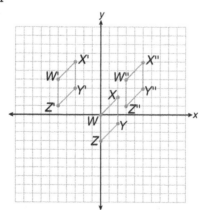

Set # 42, pages 215–217

1. $A'(-5, 1), B'(-7, 4), C'(-2, 6)$

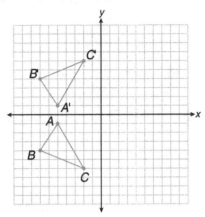

2. $J'(6, 3), K'(4, 7), L'(-2, 6)$

Triangles are congruent; same shape and size.

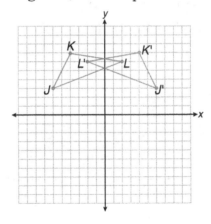

3. $J'(2,2), U'(6,2), N'(8,5), E'(4,5)$

$J''(-2,2), U''(-6,2), N''(-8,5), E''(-4,5)$

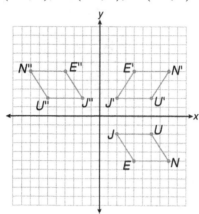

4. a. $U'(0,4), S'(-4,4), A'(-4,0)$

   b. $U''(0,-4), S''(4,-4), A''(4,0)$

   c. $U'''(0,-4), S'''(-4,-4), A'''(-4,0)$

   d. 64 square units

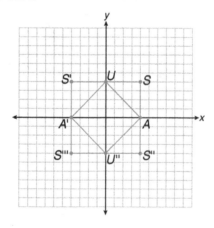

Set # 43, pages 222-223

1. a. $(-y, x)$ c. $(y, -x)$
   b. $(-x, -y)$ d. (x, y)

2.

	90°	180°	270°	360°
A	$(-4, 3)$	$(-3, -4)$	$(4, -3)$	$(3, 4)$
B	$(-2, -5)$	$(5, -2)$	$(2, 5)$	$(-5, 2)$
C	$(8, 2)$	$(-2, 8)$	$(-8, -2)$	$(2, -8)$
D	$(6, -1)$	$(1, 6)$	$(-6, 1)$	$(-1, -6)$
E	$(-7, 0)$	$(0, -7)$	$(7, 0)$	$(0, 7)$

3. $(7, -3)$

4. $(2, -6)$

5. $(-5, 1)$

6. a. See graph.

   b. $A'(6, 3), B'(2, 5), C'(6, 7)$

   c. $A''(-3, 6), B''(-5, 2), C''(-7, 6)$

   d. $A'''(-6, -3), B'''(-2, -5), C'''(-6, -7)$

   e. Congruent, same shape, same size.

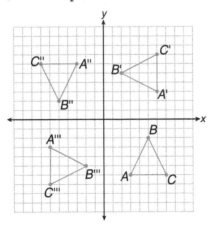

Set # 44, page 226

1. a. See graph.

   b. See graph. $L'(4, 1), M'(5, 4), N'(2, 3)$

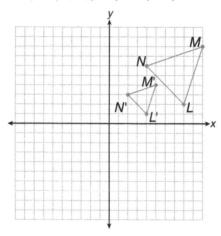

2. a. See graph.

   b. See graph. $P'(-6, 6), Q'(9, 6), R'(0, -6)$

   c. See graph. $P''(-1, 1), Q''(1.5, 1), R''(0, -1)$

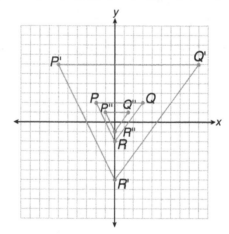

Set # 45, page 229

1. See graph.

$A'(-2, -1), B'(-3, -7), C'(2, -8), D'(1, 0)$

$A''(3, 3), B''(2, -3), C''(7, -4), D''(6, 4)$

All of these figures are congruent.

They are all the same shape and same size.

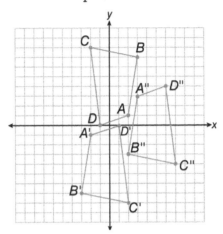

2. See graph.

$X'(2, -8), Y'(5, -10), Z'(5, -2)$

$X''(-2, -8), Y''(-5, -10), Z''(-5, -2)$

To move $Z''(-5, -2)$ to $Z'''(-1, 5)$ move 4 right and 7 up.

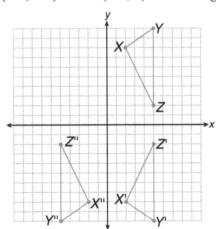

3. Triangle ABC was rotated 90° counterclockwise and then $A'B'C'$ was reflected over the x-axis to get $A''B''C''$.

Chapter 8

Real Number System

Squares and Square Roots

When you square a number, you multiply the number by itself. So
$1 \times 1 = 1$; $2 \times 2 = 4$; $3 \times 3 = 9$. The answers you get when you
square a number are called perfect squares. Let's list our perfect
squares up to 400 (20×20).

When you take the **square root** of a number, you try to figure out
what number, when multiplied by itself, will give this number. Think
of it this way: You are "undoing" the perfect square.

For example, what is the square root of 36? Ask yourself, what number times itself is 36? It is 6! Here is how we write it:

$$\sqrt{36} = 6$$

This symbol $\sqrt{}$ is called a **radical sign**.

The number under the radical is called the **radicand**.

What is the square root of 100? Ask yourself, what number, times itself, equals 100? 10!

What is the square root of 49? What number, times itself, equals 49? 7!

You need to know the perfect squares in the previous chart, because you can take the square root of these numbers. Let's make a new chart, showing all of your square roots.

PAINLESS TIP

Square Roots (Memorize these!)

$\sqrt{1} = 1$	$\sqrt{36} = 6$	$\sqrt{121} = 11$	$\sqrt{256} = 16$
$\sqrt{4} = 2$	$\sqrt{49} = 7$	$\sqrt{144} = 12$	$\sqrt{289} = 17$
$\sqrt{9} = 3$	$\sqrt{64} = 8$	$\sqrt{169} = 13$	$\sqrt{324} = 18$
$\sqrt{16} = 4$	$\sqrt{81} = 9$	$\sqrt{196} = 14$	$\sqrt{361} = 19$
$\sqrt{25} = 5$	$\sqrt{100} = 10$	$\sqrt{225} = 15$	$\sqrt{400} = 20$

To help you visualize squares and square roots, think about the relationship between the area of a square and the length of the sides of a square.

Area = 4
Side length = $\sqrt{4}$ = 2

Area = 9
Side length = $\sqrt{9}$ = 3

Every positive number has two square roots, one positive and one negative. The square root of 16 is 4 because $4(4) = 16$. The other square root of 16 is -4 because $-4(-4)$ also equals 16.

Technically, we can write $\sqrt{6} = \pm4$, meaning "plus or minus" 4.

Usually, we give only the positive answer when finding the square root. Your calculator will give you only the positive answer. However, it is important to realize that there are two answers.

As the mathematicians say, the positive answer is the **principal square root**. When you say the square root of 81 is 9, you are finding the principal square root. Unless you are asked to find two square roots, in this book we will find only the positive square roots.

1+2=3 MATH TALK!

To square a number means to multiply the number by itself.

Three squared $= 3^2 = 9$

To find the square root of a number means what number times itself will give that number?

The square root of 9 is 3 because 3 times 3 equals 9. $\sqrt{9} = 3$

BRAIN TICKLERS   Set # 46

1. List the perfect squares up to 100.

2. Find two square roots of the number 169.

3. Evaluate: 4^2

4. Evaluate: 14^2

5. Find the square root of 400.

6. A checkerboard has 64 total squares, 32 black and 32 white. How many squares are on each side of the board?

7. A square room has 225 square feet of carpet. What is the length of each side of the carpet? Will this carpet fit into a room that is 14 feet by 16 feet? Explain your reasoning.

8. A living room measures 31.8 feet by 9.6 feet. What is the largest square rug that could be put into this room? Justify your answer.

9. Can you get the $\sqrt{-400}$? Explain.

(Answers are on page 251.)

Estimating Square Roots

For everyday problems, not all measurements come out to be a perfect square, such as 25 or 36. It is important to learn to estimate square roots that are irrational numbers.

Example 1:

What whole number is the $\sqrt{88}$ closest to?

Solution:
Think of the perfect squares closest to 88.
$9^2 = 81$ and $10^2 = 100$.
Since 88 is between 81 and 100, but closer to 81, $\sqrt{88}$ is about 8.

Example 2:

The $\sqrt{28}$ lies between which two consecutive numbers?

Solution:
Again, think of your perfect squares. $\sqrt{25} = 5$ and $\sqrt{36} = 6$. The $\sqrt{28}$ is in between these two perfect squares; therefore, $\sqrt{28}$ is between 5 and 6.

Example 3:

If your floor has an area of 200 square feet, about how much square carpet do you need, in feet?

Solution:
We know that $14^2 = 196$ (too low) and $15^2 = 225$ (too high). Since 200 is between 196 and 225, you need between 14 and 15 feet of carpet. To make sure you have enough, you would need to buy 15 feet of carpet.

Example 4:

A customer has a square window with an area of 633 square inches. How long is each side of the window to the nearest inch? If the customer wanted the window divided into four equal squares, what would be the area of each of these squares to the nearest inch?

Solution:
Since the window is square, we need to estimate the square root of 633.

$$\boxed{\begin{array}{c} 633 \\ \text{sq. in.} \end{array}} \ \ ?$$

$25 \times 25 = 625$ (too low) and $26 \times 26 = 676$ (too high). Since 633 is closest to 625, we can estimate the square window to have sides that are about 25 inches on each side.

To divide the window into four equal squares, divide 625 by $4 = 156.25$

We know $12 \times 12 = 144$ and $13 \times 13 = 169$. 156.25 is closest to 144, so the side length of each little square would be about 12. The area of each of these squares would be about 144.

BRAIN TICKLERS   Set # 47

1. Between which two whole numbers does $\sqrt{80}$ lie?

2. Jackie thinks $\sqrt{29}$ is close to 6. Tim thinks it is close to 7. Who is correct and why?

Each square root is between two numbers. Name the integers.

3. $\sqrt{51}$

4. $\sqrt{110}$

5. $\sqrt{140}$

6. Steph needs to fence in her square backyard for her horse. She measures the backyard to be 925 square feet. If fencing is sold by the foot, how many feet of fencing will Steph need?

7. The Downeys are moving into a new house. Averill wants the room with the largest area. Which room should she pick? The square bedroom with a length of 15 feet, the room with an area of 196 square feet, or the room that is 12 feet by 18 feet?

(Answers are on page 251.)

Cubes and Cube Roots

We know to square a number, you multiply a number by itself two times: $3^2 = 9$.

To **cube** a number, you just multiply the number by itself three times.

$$3^3 = 3 \times 3 \times 3 = 27$$

How can we show 3 cubed?

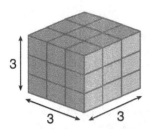

$$3 \times 3 \times 3 = 3^3 = 27$$

Other cubes:

$$2^3 = 2 \times 2 \times 2 = 8$$
$$4^3 = 4 \times 4 \times 4 = 64$$
$$5^3 = 5 \times 5 \times 5 = 125$$
$$6^3 = 6 \times 6 \times 6 = 216$$

Just as we can find the square root of a number ($\sqrt{25} = 5$), we can also find the cube root of numbers!

2 cubed = 8 because $2 \times 2 \times 2 = 8$. Now think, the cube root of 8 is what number? What can I multiply three times together to get 8, and it has to be the same number?

Two Cubed		Cube Root of 8
$2^3 = 8$	and	$\sqrt[3]{8} = 2$

This is the symbol that means **cube root**. It is a radical symbol with a little three on the outside of it to indicate cube.

$$\sqrt[3]{}$$

When you have a problem such as $\sqrt[3]{64} = 4$, you say the cube root of 64 = 4. Why? Because $4 \times 4 \times 4 = 64$ or because $4^3 = 64$.

	Cube	
1		1
2		8
3		27
4		64
5		125
6		216
7		343
	Cube root	

You can also find the cube root of negative numbers!

Notice $5 \times 5 \times 5 = 125$ (so $5^3 = 125$) and $-5 \times -5 \times -5 = -125$ so $(-5)^3 = -125$.

What is $\sqrt[3]{-8}$? The answer is -2, because $(-2)(-2)(-2) = -8$.

BRAIN TICKLERS   Set # 48

Evaluate.

1. $(-3)^3$

2. 8^3

3. $(-1)^3$

4. $\left(\dfrac{1}{2}\right)^3$

5. $\sqrt[3]{216}$

6. $\sqrt[3]{-27}$

7. Skyler thinks that since you can take the cube root of negative numbers such as $\sqrt[3]{-64}$, you should also be able to take the square root of negative numbers such as $\sqrt{-64}$. Is this true? Explain your reasoning.

(Answers are on pages 251–252.)

The Real Number System

In science class, you learn how to classify animals. For example, a gecko is a lizard, which is a reptile, which is an animal. This is an animal system: animal → reptile → lizard → gecko. In math, we classify numbers. The real number system consists of all the points on a number line.

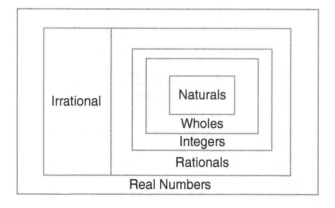

REAL NUMBER SYSTEM

Rational Numbers	Irrational Numbers
A number that can be written as a fraction	A number that cannot be written as a fraction
A decimal that ends/terminates	A decimal that does not repeat and does not end/terminate
A decimal that repeats	

Examples:

$\frac{1}{2}$ 0.125   0.333 . . .

$\sqrt{25}$ 1.25

Examples:

π 0.123456 . . . $\sqrt{46}$

Rational numbers can be broken into more specific families.

Natural numbers {1, 2, 3, 4 . . .}

The natural numbers can also be called counting numbers; they are the numbers you count.

Whole numbers {0, 1, 2, 3, 4 . . .}

The whole numbers are all of the natural numbers plus zero.

Integers {. . . −3, −2, −1, 0, 1, 2, 3, 4 . . .}

The integers are all of the natural numbers, their opposites, and zero.

To classify or name the sets of numbers that a number belongs to, you name any set that it satisfies.

PAINLESS TIP

Real Number System

Rational   Irrational
  ↓
Natural (1, 2, 3 . . .)
  ↓
Whole (0, 1, 2, 3 . . .)
  ↓
Integer (. . . −2, −1, 0, 1, 2 . . .)

Exciting examples:

For each problem, name all the sets of numbers to which each real number belongs.

1. 5

Solution:
First decide if 5 is rational or irrational. It can be written as a fraction, so it is rational. It is also a natural number, whole number, and integer.

2. −4

Solution:
−4 is rational and an integer.

3. $\sqrt{17}$

Solution:
Since you cannot find the exact value of $\sqrt{17}$, it is irrational.

4. $\sqrt{100}$

Solution:
The $\sqrt{100} = 10$, 10 is a rational number, a natural number, a whole number, and an integer.

CAUTION—Major Mistake Territory!

A number, such as 5, can belong to more than one set of numbers. Think of people. A woman can be a mother, a daughter, a sister, and an aunt. You have to name all of the sets a number belongs to. 5 is a rational number, but it is also a natural number, a whole number, and an integer.

BRAIN TICKLERS   Set # 49

Place a check mark in each appropriate column, to represent the set(s) to which the number belongs.

Example	Rational	Irrational	Integer	Whole	Natural
1. 6					
2. −2					
3. $\sqrt{25}$					
4. 1.25					
5. 0.126498503 . . .					
6. $\dfrac{3}{4}$					
7. $\sqrt{18}$					
8. 0.55555 . . .					

9. Which number in the list is rational? Explain your reasoning.

   $\sqrt{12}$ $\sqrt{13}$ 0.76

10. Which number in the list is irrational? Explain.

   $\sqrt{144}$ $\sqrt{19}$ 0.5 $\dfrac{2}{3}$

11. List three examples of integers and explain your reasoning.

12. Find a rational number between 2.16 and $\dfrac{11}{5}$.

(Answers are on page 252.)

Rational Numbers—Repeating and Nonrepeating Decimals

A rational number can be written as a fraction. Fractions expressed as decimals will either repeat or not repeat.

Examples:

Terminating	Repeating
$\dfrac{1}{2} = 0.5$	$= 0.3333333333333\dots$
$\dfrac{2}{5} = 0.4$	$= 0.18181818181818\dots$

You should already be able to write terminating decimals as fractions.

Examples:

Write each number as a fraction.

	7	0.2	0.375	4.37
Solutions:	$\dfrac{7}{1}$	$\dfrac{2}{10} = \dfrac{1}{5}$	$\dfrac{375}{1{,}000} = \dfrac{3}{8}$	$4\dfrac{37}{100}$

Repeating decimals

To read a repeating decimal, look at the following:

$0.\overline{3}$ is read as "point 3 repetend 3"

$0.\overline{18}$ is read as "point 18 repetend 18"

$0.6\overline{3}$ is read as "point 63 repetend 3"

Write the fraction as a repeating decimal using a calculator:

1. $\dfrac{2}{3} = 0.6666666\dots$

2. $\dfrac{2}{99} = 0.020202\dots$

3. $\dfrac{13}{11} = 1.181818\dots$

To convert repeating decimals to fractions (algorithm)

1. Let $N =$ the repeating decimal.

2. Multiply both sides of the equation by 10^n, in which n represents the number of digits that repeat.

3. Subtract Step 1 from Step 2.

4. Solve the resulting one-step equation.

Examples:

1. Convert $0.\overline{6}$ into a fraction.

Solution:
Let $N = 0.6666666\ldots$

Since there is only one repeating number (6), multiply both sides of the equation by $10^1 = 10$.

So $10N = 6.6666666\ldots$

Subtract N from $10N$:

$$10N = 6.6666666\ldots$$
$$-N = 0.666666\ldots$$
$$\frac{9N}{9} = \frac{6}{9}$$
$$N = \frac{2}{3}$$

2. Convert $0.\overline{28}$ into a fraction.

Solution:
Let $N = 0.28282828\ldots$

Since there are two repeating numbers (28), multiply both sides of the equation by $10^2 = 100$.

So $100N = 28.282828\ldots$

Subtract N from $100N$:

$$100N = 28.28282828\ldots$$
$$\underline{-N = 0.28282828\ldots}$$
$$\frac{99N}{99} = \frac{28}{99}$$
$$N = \frac{28}{99}$$

BRAIN TICKLERS   Set # 50

Express each decimal as a fraction in lowest terms.

1. 0.88888888 . . .

2. 0.7777777 . . .

3. 0.3636363636 . . .

4. $0.\overline{45}$

(Answers are on page 252.)

BRAIN TICKLERS—THE ANSWERS

Set # 46, page 240

1. 1, 4, 9, 16, 25, 36, 49, 64, 81, 100

2. 13, −13

3. 16

4. 196

5. 20

6. 8

7. 15 feet; No, 14 × 16 = 224 and carpet is 225 square feet.

8. 31.8(9.6) = 305.28 square feet
   Closest perfect square = 289, so 17 × 17 square rug.

9. No, we cannot multiply two of the same number (with the same sign) and get a negative number.

Set # 47, page 242

1. 8 and 9

2. Jackie is correct because 29 is closer to 36 than 49.

3. 7 and 8

4. 10 and 11

5. 11 and 12

6. Each side is approximately 31 feet. 31(4) = 124 feet of fencing.

7. Averill should pick the square room with a length of 15 feet.

Set # 48, page 244

1. −27

2. 512

3. −1

4. $\dfrac{1}{8}$

5. 6

6. -3

7. No, $(-4)(-4)(-4) = -64$, but you can't have two negative numbers that multiply to -64. For example, $(-8)(-8) = 64$.

Set # 49, page 247

Example	Rational	Irrational	Integer	Whole	Natural
1. 6	x		x	x	x
2. -2	x		x		
3. $\sqrt{25}$	x		x	x	x
4. 1.25	x				
5. 0.126498503 ...		x			
6. $\dfrac{3}{4}$	x				
7. $\sqrt{18}$		x			
8. 0.55555 ...	x				

9. 0.76, because it can be written as a fraction (76/100).

10. $\sqrt{19}$ cannot be written as a fraction.

11. $-3, 0, 1$; any positive or negative whole number.

12. Any number between 2.16 and 2.2; for example, 2.18.

Set # 50, page 250

1. $\dfrac{8}{9}$

2. $\dfrac{7}{9}$

3. $\dfrac{4}{11}$

4. $\dfrac{5}{11}$

Chapter 9

Geometry

Pythagorean Theorem

PAINLESS TIP

In any right triangle, the square of the length of the hypotenuse is equal to the sum of the squares of the lengths of the legs.

$$a^2 + b^2 = c^2$$

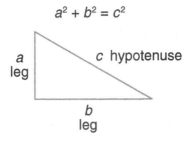

Hypotenuse

The side opposite the right angle. It is the longest side of the triangle.

Legs

The two shorter sides, which form the right angle of the triangle.

Examples:

1. Find the base of a right triangle that has a leg of 9 inches and a hypotenuse of 15 inches.

   Solution:
   Draw a picture and label your information.

   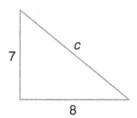

   Use the formula: $a^2 + b^2 = c^2$
   $$9^2 + b^2 = 15^2$$
   $$81 + b^2 = 225$$
   $$\underline{-81 \qquad\quad -81}$$
   $$b^2 = 144 \quad \leftarrow \text{ to undo a square,}$$
   $$\text{take the square}$$
   $$\text{root}$$

   $$b = \sqrt{144}$$
   $$b = 12$$

   The base (other leg) of the triangle is 12 inches.

2. Find the length of the hypotenuse of a right triangle if the legs measure 7 centimeters and 8 centimeters. Round to the nearest tenth.

   Solution:
   $$a^2 + b^2 = c^2$$
   $$7^2 + 8^2 = c^2$$
   $$49 + 64 = c^2$$
   $$113 = c^2$$
   $$\sqrt{113} = c$$
   $$10.6 = c$$

   The hypotenuse is 10.6 centimeters.

PAINLESS TIP

It does not matter which leg is *a* or *b*.

But the hypotenuse, or longest side, must always be *c*.

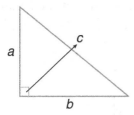 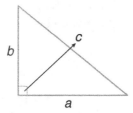

3. A helicopter flies 10 miles due north and then 28 miles due east. The helicopter then flies in a straight line back to its starting point. How many total miles did the helicopter fly? Round to the nearest whole mile.

Solution:
Draw a picture and label your information.

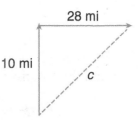

The question is asking how many **total miles** the helicopter flew, but we need to find the hypotenuse first, before we can find the total miles.

Use the formula to find the hypotenuse:

$$a^2 + b^2 = c^2$$
$$10^2 + 28^2 = c^2$$
$$100 + 784 = c^2$$
$$884 = c^2$$
$$c = \sqrt{884}$$
$$c = 29.73 \text{ (don't round until end)}$$

Now add up all the miles:
$28 + 10 + 29.73 = 67.73$, and now round to the nearest mile, so 68 miles.

The helicopter flew a total of 68 miles.

4. Find the length of the hypotenuse of the triangle with the given coordinates: $A(0, 5)$, $B(0, 1)$, and $C(3, 1)$.

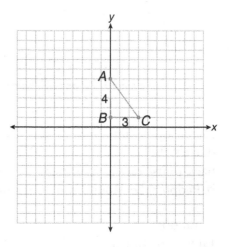

Plot coordinates on graph paper, draw the triangle, and find the lengths of the legs by counting the boxes.

$$a^2 + b^2 = c^2$$
$$3^2 + 4^2 = c^2$$
$$9 + 16 = c^2$$
$$25 = c^2$$
$$\sqrt{25} = c$$
$$5 = c$$

The hypotenuse is 5 units.

BRAIN TICKLERS   Set # 51

1. Find the length of the missing side of the right triangle. Round your answer to the nearest tenth.

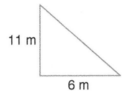

11 m

6 m

2. A right triangle has one leg that measures 15 inches and a hypotenuse that measures 17 inches. Find the measure of the second leg.

3. Mr. and Ms. Pierce commute to work separately each morning. Mr. Pierce drives 15 miles due east to work, and Ms. Pierce drives 13 miles south from their house to her office. About how many miles away do Mr. and Ms. Pierce work from each other?

4. Find the length of the hypotenuse of a triangle with the coordinates (1, −2), (1, 7), and (13, −2). Round to the nearest whole number.

5. A flagpole is 45 feet tall. A rope is tied to the top of the flagpole and secured to the ground 10 feet from the base of the flagpole. Will a 35-foot piece of rope be long enough to secure the flagpole? Justify your answer.

(Answers are on page 267.)

Volume of Cylinders

How much soda is in a soda can? How much paint is in a paint can? To find the inside measurement, you need to know the volume of a cylinder!

$$V = \pi r^2 h$$

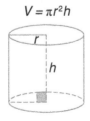

r

h

$$V = \pi r^2 h$$
Volume = (pi)(the radius squared)(height)

Examples:

1. Find the volume of a right cylinder with a radius of 2 centimeters and a height of 7 centimeters. Express your answer in terms of π and rounded to the nearest hundredth.

Solution: $V = \pi r^2 h$
$V = \pi(2^2)(7)$
$V = \pi(4)(7)$
$V = 28\pi$ (answer in terms of π)

$28\pi = 87.9645943 = 87.96$ cm³

2. Which can of pizza sauce is a better buy, if both cans are cylinders? Can A with a diameter of 10 centimeters and a height of 20 centimeters for $3.49 or Can B that has a diameter of 8 centimeters and a height of 10.5 centimeters for $1.09? Justify your answer.

Solution:
Find the volume of the two cylinders (cans).

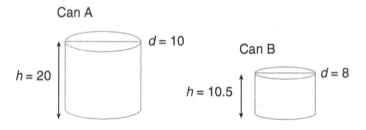

Can A
$d = 10$
$h = 20$

Can B
$d = 8$
$h = 10.5$

(Notice diameter is given, but you need radius for the formula.)

$V = \pi r^2 h$ $V = \pi r^2 h$
$V = \pi(5^2)(20)$ $V = \pi(4^2)(10.5)$
$V = \pi(25)(20)$ $V = \pi(16)(10.5)$
$V = 500\pi$ $V = 168\pi$

500π costs $3.49. 168π costs $1.09.

How can we see which is a better buy?

Can A holds 500π cubic centimeters of sauce. This is 1,570.796327 cubic centimeters for $3.49.

Can B holds 527.787566 cubic centimeters for $1.09. Notice this is about one-third of the other can. If you bought three cans of Can B, you would have a volume of 1,583.362698 cubic centimeters for $3.27.

You are getting more sauce in Can B for a cheaper price! Can B is a better buy!

BRAIN TICKLERS   Set # 52

1. Find the volume of a cylinder with a height of 11 centimeters and a radius of 1.5 centimeters. Round your answer to the nearest tenth.

2. Grain is stored in cylindrical structures called silos. Find the volume of a silo with a diameter of 12 feet and a height of 20 feet. Round to the nearest tenth.

3. A cylinder has a height of 5 inches and radius of 2 inches. Explain whether tripling the height of the cylinder will triple the volume of the cylinder. Justify your answer.

4. Which cylindrical vase will hold more water? A vase that is 10 inches tall with a diameter of 15 inches or a vase that is 15 inches tall and has a diameter of 10 inches? Justify your answer.

5. Cole is making candles. His cylindrical mold is 9 inches tall and has a base with a diameter of 3 inches. If he has an existing candle made that is 56.5 cubic inches, did it come from this candle mold? Explain your reasoning.

6. A student had to find the volume of the following can. Explain the error(s) the student made in the work shown below.

$V = \pi r^2 h$
$V = \pi (10^2)(16)$
$V = \pi (20)(16)$
$V = 320\pi$
$V = 1005.31$

10 cm

16 cm

(Answers are on page 267.)

Volume of Cones

Did you know that three cones will fit into a cylinder? If you already know the cylinder formula $V = \pi r^2 h$, then to find one cone, you take one-third of the cylinder formula!

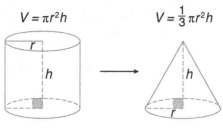

$$V = \pi r^2 h \qquad\qquad V = \frac{1}{3}\pi r^2 h$$

$$\text{Volume of a Cone} = V = \frac{1}{3}\pi r^2 h$$

Volume = (one-third)(pi)(radius squared)(height)

Examples:

1. Find the volume of a cone with a radius of 4 centimeters and a height of 8 centimeters. Express your answer in terms of π and rounded to the nearest tenth.

Solution:

$$V = \frac{1}{3}\pi r^2 h$$

$$V = \frac{1}{3}\pi (4^2)(8)$$

$$V = \frac{1}{3}\pi (16)(8)$$

$$V = \frac{128}{3}\pi \quad \text{(answer in terms of } \pi)$$

$$\frac{128}{3}\pi = 134.0412866 = 134.0 \text{ cm}^3$$

2. Find the volume of the given cone.
   Leave in terms of π.

   Solution:
   In this problem, you are given a radius
   of 12 millimeters and a slant height of
   13 millimeters. We need the height of
   the cone for the volume formula.

   Since we have a right triangle, you
   can find the height using the
   Pythagorean Theorem!

$$12^2 + h^2 = 13^2$$
$$144 + h^2 = 169$$
$$25 = h^2$$
$$5 = h$$

   Now find the volume:

$$V = \frac{1}{3}\pi r^2 h$$

$$V = \frac{1}{3}\pi(12^2)(5)$$

$$V = \frac{1}{3}\pi(144)(5)$$

$$V = \frac{1}{3}\pi(720)$$

$$V = 240\pi$$

   The volume of the cone is 240π cubic millimeters.

3. Which would give you a bigger volume? Doubling the
radius or doubling the height? Justify your answer using
the cone given below.

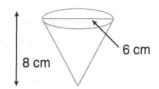

8 cm

6 cm

Solution:

Solve the problem both ways.

Doubling the radius	Doubling the height
$V = \frac{1}{3}\pi r^2 h$	$V = \frac{1}{3}\pi r^2 h$
$V = \frac{1}{3}\pi(6^2)(8)$	$V = \frac{1}{3}\pi(3^2)(16)$
$V = \frac{1}{3}\pi(36)(8)$	$V = \frac{1}{3}\pi(9)(16)$
$V = \frac{1}{3}\pi(288)$	$V = \frac{1}{3}\pi(144)$
$V = 96\pi$	$V = 48\pi$

Doubling the radius will give you a greater volume.

BRAIN TICKLERS   Set # 53

1. Find the volume of a cone with a radius of 3.1 feet and a height of 6.7 feet. Round your answer to the nearest tenth.

2. Find the volume of a cone with a diameter of 10 inches and a height of 5.2 inches. Round your answer to the nearest hundredth.

3. Julie made mini waffle cones for a birthday party. Each waffle cone was 3 inches high and had a radius of 0.75 inches. Will Julie's waffle cones hold 2 cubic inches of ice cream? Justify your reasoning.

4. You have a construction cone that has a height of 3 feet and a diameter of 8 inches. How would the volume of the cone change if you doubled the height? Doubled the radius? Show your work. Explain your answers.

5. An oil funnel is in the shape of a cone. The cone has a diameter of 3 inches and a height of 5 inches. If the end of the funnel is plugged, how much oil can the cone hold before it overflows? Round your answer to the nearest tenth. Can the funnel hold 11 cubic inches of oil? Justify your reasoning.

(Answers are on page 267.)

Volume of Spheres

Think of different sports or games that involve a ball. Basketball, baseball, tennis, and soccer are just a few. The mathematical name for the shape of a ball is **sphere**.

A **sphere** is the set of points in three dimensions that are a fixed distance from a given point, also known as the center. If you were to cut the sphere through its center (divide it into two halves), you get a **hemisphere**.

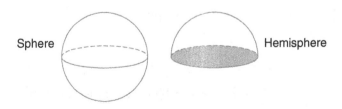

Sphere   Hemisphere

PAINLESS TIP

Volume Formulas

Volume of a sphere: $\frac{4}{3}\pi$ times the cube of the radius, r

Volume of a sphere: $V = \frac{4}{3}\pi r^3$

Examples:

1. Find the volume of a sphere with a radius of 9 inches. Round your answer to the nearest cubic foot.

   Solution:

   $$V = \frac{4}{3}\pi r^3$$

   $$V = \frac{4}{3}\pi (9/12)^3$$

   $$V = 1.767$$

   $$V = 2 \text{ cubic feet}$$

2. Find the volume of a sphere with a diameter of 20 centimeters. Leave your answer in terms of π and rounded to the nearest tenth.

   Solution:

   $$V = \frac{4}{3}\pi r^3$$

   $$V = \frac{4}{3}\pi (10)^3$$

   $$V = \frac{4}{3}\pi (1,000)$$

   $$V = \frac{4,000}{3}\pi \text{ (in terms of } \pi)$$

   $$V = 4,188.8 \text{ cm}^3 \text{ (rounded answer)}$$

3. Explain why the formula for the volume of a hemisphere is $V = \dfrac{2}{3}\pi r^3$. How can you use the volume of a sphere formula to also get the answer?

Solution:
The volume of a sphere is $V = \dfrac{4}{3}\pi r^3$. A hemisphere is half of a sphere. Take half of four-thirds to find the formula for a hemisphere.

$$\frac{1}{2} \times \frac{4}{3} = \frac{4}{6}, \text{which is equal to} \frac{2}{3}$$

So the formula for a hemisphere would be $V = \dfrac{2}{3}\pi r^3$.

Or just use the sphere formula and divide your answer by 2.

$$V = \left(\frac{4}{3}\pi r^3\right) \div 2$$

4. A 3-inch-high ice cream cone is packed full of ice cream, and a hemisphere of ice cream is placed on top. If the diameter of the cone is 5 inches, find the volume of the cone, including the ice cream on top. Round your answer to the nearest hundredth.

Solution:
You need the volume of the cone + the volume of the hemisphere.

$$V = V_c = \frac{1}{3}\pi r^2 h + V_b = \left(\frac{4}{3}\pi r^3\right) \div 2$$
$$V = \frac{1}{3}(\pi)(2.5)^2(3) + \left(\frac{4}{3}\pi(2.5)^3\right) \div 2$$
$$V = 6.25\pi + \frac{1}{2}(65.44984695)$$
$$V = 52.36 \text{ in}^3$$

BRAIN TICKLERS   Set # 54

1. A snow globe has a diameter of 4 inches. Find the volume of the globe to the nearest thousandth.

2. A beach ball is spherical in shape. Find the volume of the beach ball to the nearest tenth if the radius is 3 inches.

3. Becca thinks that if you double the radius of a sphere, it will double the volume. Is she right? Justify your reasoning.

4. Find the volume of a hemisphere with a diameter of 18 inches to the nearest hundredth.

5. A water balloon has a radius of 5 inches. Can the water balloon hold 525 cubic inches of water or will it pop? Explain your reasoning.

6. Epcot Center's Spaceship Earth is shaped like a sphere. It has a diameter of 160 feet. Is the volume of the attraction more or less than 1 million cubic feet? By much?

(Answers are on page 268.)

BRAIN TICKLERS—THE ANSWERS
Set # 51, page 257

1. 12.5 meters

2. 8 inches

3. $19.8 \approx 20$ miles

4. 15

5. No, because $\sqrt{2,125} \approx 46$ feet. A 35-foot piece of rope is not long enough.

Set # 52, page 259

1. 77.8 cubic centimeters

2. 2,261.9 cubic feet

3. Yes, the original $V = 20\pi$ and the new $V = 60\pi$.

4. A vase that is 10 inches tall holds more water (562.5π vs. 375π).

5. No, a mold has a volume of 63.617 cubic inches; the existing candle is smaller.

6. Used diameter of 10, instead of radius of 5. 10^2 does not $= 20$; it equals 100.

Set # 53, page 263

1. 67.4 cubic feet

2. 136.14 cubic inches

3. No, the volume of the cone is only 1.767 cubic inches.

4. The volume of the original cone is $V = 192\pi$ cubic inches (0.349 cubic feet). Doubling the height $= 384\pi$ cubic inches (0.698 cubic feet), which is double the original volume. Doubling the radius $= 768\pi$ cubic inches (1.396 cubic feet), which is 4 times the original volume.

5. Yes, the volume of the funnel is 11.8 cubic inches, so 11 cubic inches will fit.

Set # 54, page 266

1. 33.510 cubic inches

2. 113.1 cubic inches

3. No, the volume is eight times larger because two gets cubed.

4. 1,526.81 cubic inches

5. No, it will pop. The balloon has a volume of 523.598 cubic inches, which is smaller than 525 cubic inches.

6. The volume is more than 1 million (2,144,660.585 cubic feet). The difference is 1,144,660.585 cubic feet.

Chapter 10

Introduction to Bivariate Statistics

Two-Way Tables

There are all types of data or information we can collect about people, sports, jobs, etc. Data such as hair color, favorite sport, and favorite food are examples of data that cannot be measured numerically (with a number). These are "categories," and this type of data is called **categorical or qualitative data**.

Data that involves variables that can be measured numerically, such as height, weight, shoe size, and number of at bats, are all examples of **quantitative data**.

1+2=3 **MATH TALK!**

Qualitative: think "quality" or characteristic

Quantitative: think "quantity" or number

The type of information you are collecting will determine if you want to make a table or a graph. In this chapter, we will review how to organize categorical data by using two-way tables.

Information about people who are surveyed can be captured in a two-way table. A **two-way table** shows data from one sample group as it relates to two different categories/categorical data. The **frequency** or number of times each value occurs will be listed in the table.

Example 1:

Trevor surveys students in his school who play sports and asks them which sport they prefer. The responses are shown in the two-way table below.

	Preferred sport			
	Basketball	Lacrosse	Soccer	Baseball/Softball
Male	25	36	48	16
Female	24	19	64	33

a. Why is this bivariate data? Is it qualitative or quantitative?
The data is bivariate because we are comparing two sets of data: people and sports. The data is qualitative because male/female and types of sports are categories and cannot be measured numerically.

b. How many males play sports? How many females?
For males, add across the table: $25 + 36 + 48 + 16 = 125$
For females, add across the table: $24 + 19 + 64 + 33 = 140$

c. How many students were surveyed?
Add all males and females: $125 + 140 = 265$

d. How many students play basketball?
Add the column: $25 + 24 = 49$

e. Out of the females surveyed, what percent of them play soccer? Round to the nearest percent.
$\dfrac{64}{140} = 0.457 = 46\%$

f. What percent of the soccer players are female?
$\dfrac{64}{112} = 0.571 = 57\%$

g. What percent of the males play baseball/softball?
$\dfrac{16}{125} = 0.128 = 13\%$

PAINLESS TIP

The way the question is asked will help you determine if you need to total the numbers up and down (columns) or left to right (rows).

Example 2:

Mr. Meyer surveyed students at his school. He found that 80 students own a cell phone and 55 of those students also own an iPad. There are 13 students who do not own a cell phone but do own an iPad. Ten students do not own either device. Construct a two-way table summarizing the data.

Solution:

Step 1: Create a table using the two categories: Cell Phone and iPad. You need to include cell phone, no cell phone, iPad, no iPad, and your totals.

Note: Not all tables show the total columns, but they are nice to have. If the total column is not provided, total the columns and rows yourself.

	iPad	No iPad	Total
Cell phone			
No cell phone			
Total			

Step 2: Use the information given to fill in the table. Read carefully!

	iPad	No iPad	Total
Cell phone	55		80
No cell phone	13	10	
Total			

Step 3: Complete the rest of the table based on the information you already know. Remember, the totals are for each row and column. The column labeled "Total" should have the same sum as the row labeled "Total."

	iPad	No iPad	Total
Cell phone	55	25	80
No cell phone	13	10	23
Total	68	35	103

Example 3:

There are 200 children signed up for a local summer camp. Of those students, 74 signed up for sports. A total of 80 children signed up for crafts, and 28 of them also signed up for sports. Construct a two-way table summarizing the data.

Step 1: Create your table, and fill in the information you know.

	Crafts	No crafts	Total
Sports	28		74
No sports			
Total	80		200

Step 2: Complete the rest of the table.

	Crafts	No crafts	Total
Sports	28	46	74
No sports	52	74	126
Total	80	120	200

A two-way table can also show relative frequencies. **Relative frequency is the ratio of the value of a subtotal to the value of the total.** A two-way table can show relative frequencies for rows or for columns rather than the actual values.

Using the table from Example 3, a table showing the relative frequencies by row and a table showing the relative frequencies by column are shown below. Values are rounded to the nearest hundredth when necessary.

> **1+2=3 MATH TALK!**
>
> To find the relative frequencies **by row**, write the ratios of each value to the total in that row.
>
> To find the relative frequencies **by column**, use the total of the columns when writing the ratios.

Relative frequencies by row

	Crafts	No crafts	Total
Sports	$\frac{28}{74} = 0.38$	$\frac{46}{74} = 0.62$	$\frac{74}{74} = 1$
No sports	$\frac{52}{126} = 0.41$	$\frac{74}{126} = 0.59$	$\frac{126}{126} = 1$
Total	$\frac{80}{200} = 0.40$	$\frac{120}{200} = 0.60$	$\frac{200}{200} = 1$

What is the relative frequency of students who want to play a sport and do a craft to all of the students who play a sport at camp?

Solution: $\frac{28}{74} = 0.38$

Relative frequencies by column

	Crafts	No crafts	Total
Sports	$\frac{28}{80} = 0.35$	$\frac{46}{120} = 0.38$	$\frac{74}{200} = 0.37$
No sports	$\frac{52}{80} = 0.65$	$\frac{74}{120} = 0.62$	$\frac{126}{200} = 0.63$
Total	$\frac{80}{80} = 1$	$\frac{120}{120} = 1$	$\frac{200}{200} = 1$

What is the relative frequency of students who want no sport and no craft at the camp out of all students who do not want to do crafts?

Solution: $\dfrac{74}{120} = 0.62$

BRAIN TICKLERS   Set # 55

1. Jordan surveyed the students in her grade about the ways they get to school in the morning. Her results are shown in the table below.

	Walk	Ride bus	Ride in car
Boys	10	25	15
Girls	16	30	19

a. How many boys were surveyed?

b. How many students were surveyed?

c. How many students ride the bus to school?

d. From all the boys surveyed, what percent of the boys walk to school?

e. Out of the girls, what percent of the girls ride the bus? Round to the nearest percent.

2. A survey was given to all 10th, 11th, and 12th graders in the local high school to see how many students were working or not working after school. Some of the data is recorded in the table below.

	Job	No job	Total
10th	42		80
11th		35	
12th	83	27	
Total			280

a. Based on the information provided, complete the rest of the table.

b. How many 11th graders have a job after school?

c. How many students do not have a job?

d. What percent (to nearest tenth) of the students surveyed have a job?

e. How many 10th graders were surveyed?

f. What is the relative frequency of 12th graders who have a job to the total number of 12th graders surveyed? Round to the nearest hundredth.

g. What is the relative frequency of 10th graders who do not have a job to the total number of students who do not have a job?

3. One hundred kids and adults were surveyed about their favorite food. Fifteen kids said they like only pizza. Twenty-six adults said they like burgers out of the 61 people who like burgers. Using this information, make a two-way table. How many kids like burgers?

4. The following data was collected as people walked into a movie theater.

	Popcorn	No popcorn	Total
Drink	75	18	93
No drink	10	8	18
Total	85	26	111

a. Find the relative frequencies by row. Round to the nearest hundredth.

b. Find the relative frequencies by column. Round to the nearest hundredth.

(Answers are on pages 289–290.)

Scatterplots

Have you ever thought about the relationship between the number of miles you drive and the gallons of gas you use? The heights of basketball players and number of points scored per game?

If we want to examine the relationship between two sets of numerical data, such as height versus shoe size, we are looking at **bivariate data.**

Univariate Data → one set of data
(e.g., age)
Bivariate Data → data for two variables
(e.g., miles and gallons)

Bivariate data deals with relationships, comparisons, or causes between two sets of numerical data. To compare the data, you can make a scatterplot.

A **scatterplot** is a plot on the coordinate plane used to compare two sets of quantitative data and to look for a correlation between those two data sets.

Example 1:

Create a scatterplot for the data shown below. The data below shows the grams of fat versus the number of calories in different types of sub sandwiches.

Fat (g)	2	4	16	20	6	24	22	8	12
Calories	180	200	420	510	280	479	679	300	400

Solution:
Let the *x*-axis represent the grams of fat and the *y*-axis represent the number of calories. Then graph each ordered pair. Easy!

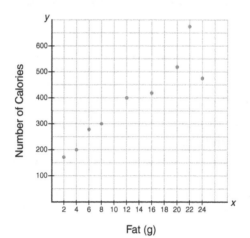

A scatterplot provides a great visual so we can look at and analyze the distribution of points or the pattern the points may or may not follow. We are looking at what is called the **correlation.**

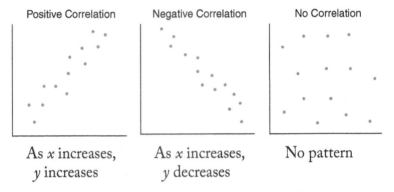

Positive Correlation	Negative Correlation	No Correlation
As *x* increases, *y* increases	As *x* increases, *y* decreases	No pattern

We also can tell from the points if the correlation is **linear** or **nonlinear.** If it is linear, the points will lie close to a line. If it is nonlinear, the points will be in the shape of a curve.

Linear	Nonlinear

Example 2:

The table below shows the amount of time students studied for a test and their test score.

Time (min)	5	10	15	20	25	30	35
Test score	55	65	65	72	78	88	88

a. Describe the correlation.

b. Is it linear or nonlinear?

Solution:

Make a scatterplot of the data.

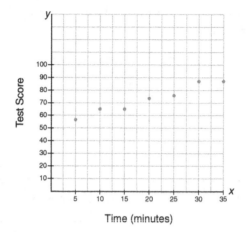

Time (minutes)

a. The scatterplot is showing that as x increases, y increases. This means that as the number of minutes a person studied increases, the test score also increases.

b. The points seem to follow a line; therefore, it is a linear relationship.

Example 3:

The following table shows the height of a ball thrown in the air versus time.

Time (sec)	0	1	2	3	4
Height (cm)	0	1,500	2,000	1,500	0

Is this linear or nonlinear? Explain.

Solution:
Nonlinear. When you make a scatterplot, the points form a curve.

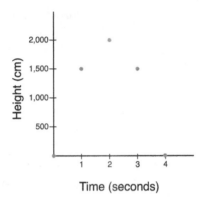

Time (seconds)

Example 4:

Make a scatterplot that shows each of the following: positive, negative, and no correlation. Label the *x*- and *y*-axis with an appropriate situation to match the data.

Solution:
Descriptions will vary, but the graphs should be similar in pattern.

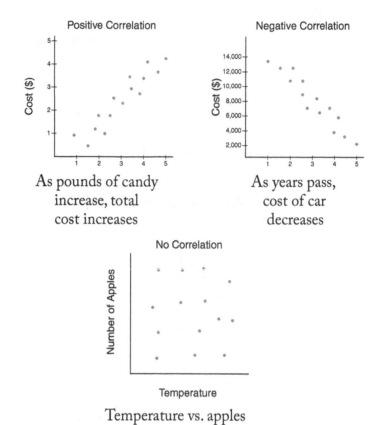

Positive Correlation

As pounds of candy
increase, total
cost increases

Negative Correlation

As years pass,
cost of car
decreases

No Correlation

Temperature vs. apples

BRAIN TICKLERS   Set # 56

1. The table below shows the population in a rural town in Iowa every ten years from 1960 to 2010.

Year	Population
1960	10,795
1970	19,177
1980	24,439
1990	30,782
2000	39,604
2010	49,045

a. Construct a scatterplot of the data.

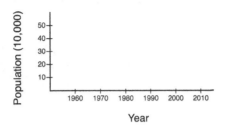

b. Interpret the shape of the scatterplot based on the distribution.

c. Based on your graph, what could the population have been in 1975?

2. The table below shows the distance traveled in miles compared to the cost for airline tickets.

Distance (miles)	500	300	612	409	100	900	200	800
Cost ($)	150	124	143	130	125	200	100	175

a. Construct a scatterplot.

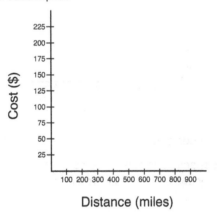

b. Describe the correlation.

c. Add to the graph two more data values that fit the correlation you described in part b.

3. The data below was recorded by a college professor. He wanted to see if there was a correlation between exam grades and the number of classes that students missed.

Missed classes	Exam grade	Missed classes	Exam grade
2	98	3	96
2	90	4	85
24	53	18	40
22	20	12	68
6	83	10	91
10	60	17	60

a. Make a prediction about the correlation.

b. Make a scatterplot of the data.

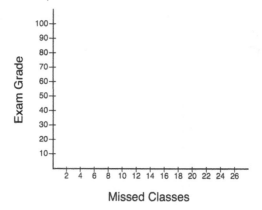

c. Describe the correlation.

d. If this relationship exists, make a conjecture about the exam grade you might receive if you miss five classes.

(Answers are on pages 290–291.)

Line of Best Fit

When collecting real-life data, there is a good chance the data will not form a perfect straight line. To help us make predictions about what may happen based on the pattern that we see, we can find what is called the **line of best fit**. This may also be called a **trend line**, **prediction line**, or a **linear regression**.

Using what you know about writing the equations of lines given two points will come in very handy for finding the line of best fit.

Example 1:

The data below shows the estimated cost of building a house in Victor, New York, since 2000.

Year	2001	2002	2003	2004	2005	2006	2007
Cost (in thousands)	170	184	187	192	230	265	257

a. Construct a scatterplot of the data, and find the line of best fit.

b. Use the line of best fit to make a prediction about what the cost of building a house in Victor was in 2012.

Steps:

1. Find two points that you think will be on the best fit line (or close to it).

2. Find the slope between these two points.

3. Write the equation of the line.

4. Use this equation to help you make other predictions about this data.

Solution:
a. Make a scatterplot:

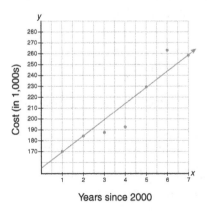

Years since 2000

Notice the x-axis is labeled years since 2000.

Choose two points that seem like they will make the best line for most of the data: $(1, 170)$ and $(7, 257)$

Find the slope between these two points:

$$\frac{y_2 - y_1}{x_2 - x_1} = \frac{257 - 170}{7 - 1} = \frac{87}{6} = 14.5$$

Find the equation of the line:

$$y = mx + b$$
$$170 = 14.5(1) + b$$
$$170 = 14.5 + b$$
$$b = 155.5$$ So the equation of the line of best fit is $y = 14.5x + 155.5$ (shown on graph).

b. Use this line to help you make predictions. Since 2012 is 12 years since 2000, use $x = 12$ in your equation.

$$y = 14.5x + 155.5$$
$$y = 14.5(12) + 155.5$$
$$y = 174 + 155.5$$
$$y = 329.5$$

If this trend continued, the cost for building a house in 2012 would have been about $329,500.

Example 2:

The following data shows the number of hours people spend on a screen (phone, laptop, other connected device, or television) since 2009.

Year	2010	2011	2012	2013	2014	2015
Time (hours/day)	7.4	8	9	9.4	9.6	10.2

a. Make a scatterplot, and find the equation for the line of best fit.

b. Interpret the slope and y-intercept.

c. Make a prediction about what the number of hours of screen time was in 2018.

Solution:

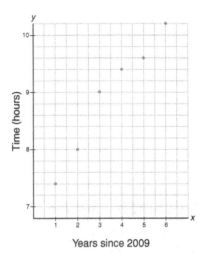

Years since 2009

a. Make a scatterplot.

Choose two points—$(1, 7.4)$ and $(5, 9.6)$—to write your equation.

$$\text{Slope} = \frac{y_2 - y_1}{x_2 - x_1} = \frac{9.6 - 7.4}{5 - 1} = \frac{2.2}{4} = 0.55$$

Find the equation of the line:

$$y = mx + b$$
$$9.6 = 0.55(5) + b$$
$$9.6 = 2.75 + b$$
$$b = 6.85$$

So the equation of the line of best fit is $y = 0.55x + 6.85$.

b. The slope $= 0.55$ means that the number hours of screen time per year is increasing by about 0.5 a year or 30 minutes a year.

The y-intercept, 6.85, means that in 2009, the starting screen time was about 6.85 or 7 hours.

c. Since 2018 is 9 years after 2009, use $x = 9$ in your equation.

$$y = 0.55x + 6.85$$
$$y = 0.55(9) + 6.85$$
$$y = 4.95 + 6.85$$
$$y = 11.8$$

If this trend continued, the screen time in 2018 would have been 11.8 hours.

BRAIN TICKLERS   Set # 57

1. The following table shows the height of basketball players in centimeters and the number of points scored during a game.

Height (cm)	Points scored
173	4
176	8
177	6
185	10
185	14
185	16
186	12
188	8
190	16
195	18
200	18
200	20
201	22
205	19

a. Make a scatterplot of the data.

b. Describe the correlation.

c. Find the line of best fit using the data points (173, 4) and (195, 18). Round to the nearest hundredth.

d. Use the equation to make a prediction about how many points might be scored if the player was 179 centimeters. Round to the nearest point.

2. The table below shows data of student GPA and the number of hours they played video games during the semester.

GPA	1.0	1.2	1.3	2.0	2.2	2.6	3	3.6	3.9	4.0
Time spent gaming (hours)	20	17	15	10	8	8	5	3	1	0

a. Make a scatterplot of the data.

b. Describe the correlation.

c. Find the equation of the line of best fit using the data points (1.2, 17) and (3.6, 3). Round to the nearest hundredth.

d. Interpret the slope.

e. Use the equation to make a prediction about a student's GPA if they spent 14 hours playing video games during the semester. Round to the nearest tenth.

3. The scatterplot below shows the temperature (°C) versus the amount of money spent on electric bills.

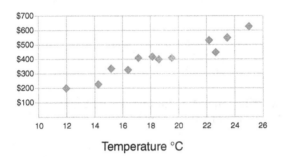

Temperature °C

a. Write an equation in slope-intercept form for the line of best fit using the points (12, 200) and (22, 510).

b. Interpret the slope.

c. When the temperature is 10°C, what is the cost?

d. Based on your equation, predict the cost of an electric bill if the temperature is 21°C.

e. Is a bill of $600 reasonable for a temperature of 13°C? Explain your reasoning. Justify using the graph and the equations.

4. Make a scatterplot of the following data values.

x	0	1	2	3	4	5	6	7
y	10	50	63	79	90	80	60	50

  a. Interpret the scatterplot based on the shape of the data.

  b. Can you find the line of best fit for this data? Explain.

5. a. On the graph below, graph the equation of the line $y = \frac{2}{3}x + 2$.

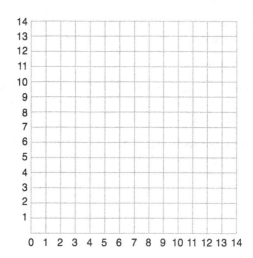

  b. On the graph, plot ten coordinates so that $y = \frac{2}{3}x + 2$ could be the line of best fit for the data.

(Answers are on pages 292–293.)

BRAIN TICKLERS—THE ANSWERS

Set # 55, pages 274-275

1. a. 50

   b. 115

   c. 55

   d. 20%

   e. 46%

2. a.

	Job	No job	Total
10th	42	38	80
11th	55	35	90
12th	83	27	110
Total	180	100	280

   b. 55

   c. 100

   d. 64.3%

   e. 80

   f. $\dfrac{83}{110} = 0.75$

   g. $\dfrac{38}{100} = 0.38$

3.

	Pizza	Burgers	Total
Kids	15	35	50
Adults	24	26	50
Total	39	61	100

Thirty-five kids like burgers.

4. a.

	Popcorn	No popcorn	Total
Drink	$\dfrac{75}{93} = 0.81$	$\dfrac{18}{93} = 0.19$	$\dfrac{93}{93} = 1$
No drink	$\dfrac{10}{18} = 0.56$	$\dfrac{8}{18} = 0.44$	$\dfrac{18}{18} = 1$
Total	$\dfrac{85}{111} = 0.77$	$\dfrac{26}{111} = 0.23$	$\dfrac{111}{111} = 1$

b.

	Popcorn	No popcorn	Total
Drink	$\dfrac{75}{85} = 0.88$	$\dfrac{18}{26} = 0.69$	$\dfrac{93}{111} = 0.84$
No drink	$\dfrac{10}{85} = 0.12$	$\dfrac{8}{26} = 0.31$	$\dfrac{18}{111} = 0.16$
Total	$\dfrac{85}{85} = 1$	$\dfrac{26}{26} = 1$	$\dfrac{111}{111} = 1$

Set # 56, pages 280–282

1. a. See graph.

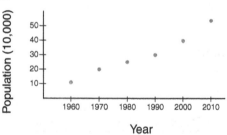

b. As years increase, population increases. It's a positive linear correlation.

c. ≈ 23,000

2. a. See graph.

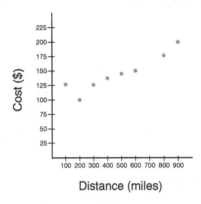

Distance (miles)

b. As distance increases, cost increases. It's a positive linear correlation.

c. Answers will vary. Points should match general pattern.

3. a. Prediction: As the number of missed classes increases, the exam grade decreases.

b. See graph.

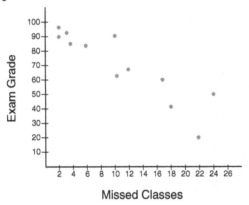

Missed Classes

c. As the number of missed classes increases, the exam grade decreases. Negative linear correlation.

d. If I miss five classes, my grade might be an 84.

Set # 57, pages 286-288

1. a. See graph.

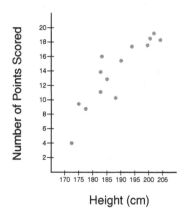

b. Positive linear correlation. As height increases, points scored increases.

c. $y = 0.64x - 106.09$

d. Player might make 8 points.

2. a. See graph.

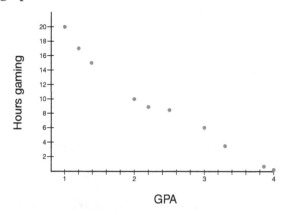

b. As GPA increases, hours of gaming decreases. It is a negative linear correlation.

c. $y = -5.83x + 24.00$

d. For every 5.83 hours you play video games, your GPA goes down by 1.

e. 1.7

3. a. $y = 31x - 172$

   b. For every $31 cost increase, the temperature increases by 1°C.

   c. $138

   d. $479

   e. No, it is too high based on the pattern of the graph. Using the line of best fit equation, a bill of $600 would be appropriate for a temperature of about 25°C.

4. a. See graph, which is nonlinear.

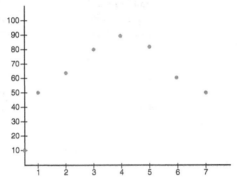

   b. No, because this is a curve.

5. a. See graph.

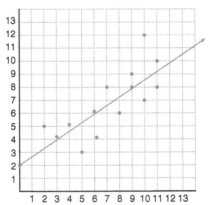

   b. Choose any 10 coordinates near the line of best fit, both above and below the line: for example $(2, 5), (5, 3), (7, 8),$ $(9, 8), (10, 12), (4, 5), (6, 4), (9, 9), (11, 8), (11, 10)$.

Index